Radar Essentials

A Concise Handbook for Radar Design and Performance Analysis

- Principles
- Equations
- Data

G. Richard Curry

PUBLISHING, INC.

Raleigh, NC
scitechpub.com

Published by SciTech Publishing, Inc.
911 Paverstone Drive, Suite B
Raleigh, NC 27615
(919) 847-2434, fax (919) 847-2568
scitechpublishing.com

Editor: Dudley R. Kay
Production Manager: Robert Lawless
Typesetting: MPS Limited, a Macmillan Company
Cover Design: Brent Beckley
Printer: Docusource

This book is available at special quantity discounts to use as premiums and sales promotions, or for use in corporate training programs. For more information and quotes, please contact the publisher.

Printed in the United States of America

10 9 8 7 6 5 4 3 2 1

ISBN: 978-1-61353-007-8

To Nancy

My companion, best friend,
and wife for 44 years

Contents

Preface

I close my radar courses and tutorials with a 35-point summary. I tell the students that if they know these "Key Radar Facts", or have them handy, they can contribute productively to any radar discussion. These points, expanded by adding tables and charts with key radar data, were published as the *Pocket Radar Guide* [1].

The next step was to meet the need for a compendium of radar information essential for design and performance analysis. It should combine elements of a radar text and a handbook. It should be comprehensive to serve as a complete radar reference, yet compact enough to fit on an engineer's desk or travel with him. It should contain the data and equations most useful to practicing radar engineers, yet also material to help radar non-experts understand and use the information.

This is that book. It contains basic principles of radar design and analysis, characteristics of the major radar components, key radar equations, and tables and charts with the most-used radar performance data. The reader can go directly to the topic of interest, where references are provided to other helpful sections. It is intended as a reference for radar and aerospace engineers and system analysts, and provides a handy desktop source as well as an essential traveling companion.

Detailed discussions, derivations, examples, and design details beyond the scope of this book can be found in the new *Principles of Modern Radar* [2], the latest edition of *Radar Handbook* [3], and *Airborne Radar* [4]. Useful analysis techniques are given in *Radar System Performance* Modeling [5], and *Radar System Analysis and Modeling* [6]. (Many of the figures in this book were produced using the custom radar functions for Excel spreadsheets provided with [5]). Radar terms are defined in *IEEE Standard Radar Definitions* [7]. Additional references to specific areas are given in the text.

I would like to acknowledge helpful discussions, encouragement, and support of Dudley Kay, President of Scitech Publishing , Inc., and the many constructive suggestions from John Milan and the other reviewers of this book.

1 | Radar Basics

1.1 Radar Concept and Operation

The word RADAR is an acronym for RAdio Detection And Ranging.

Radar employs electromagnetic propagation of energy directed toward and reflected from targets. Electromagnetic energy propagates in the atmosphere and free space in a straight line with propagation velocity $c = 3 \times 10^8$ m/s, (Sec. 6.2), except for:

- Reflection from metallic objects and those having a dielectric constant different than that of free space. Reflection from targets is the basis for radar operation (Sec. 3.1); reflection from terrain and other objects may interfere with radar operation (Sec. 4.5 and 4.7).
- Refraction, the bending of the propagation path due to differences in the propagation velocity. Refraction due to the atmosphere and the ionosphere are common radar effects (Sec. 4.3 and 4.6).
- Diffraction, the bending of the propagation path around the edge of an object. This effect is usually not significant at radar frequencies.

In the basic radar concept (Fig. 1.1), electromagnetic energy is generated by a transmitter and radiated by the transmitting antenna in the direction of the target. Some of the energy reflected by the target is collected by the receiving antenna and processed in the receiver to produce information about the target. This can include:

- Target presence, indicated by a signal return larger than the background.
- Target range, found from the round-trip propagation time, t. For monostatic radars, (Sec. 1.3), the target range, R is given by

$$R = \frac{ct}{2} \tag{1.1}$$

1

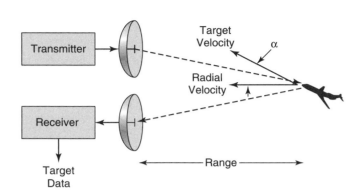

Fig. 1.1   Basic radar concept

- Target radial velocity V_R, the component of the target velocity, V, in the direction of the radar for monostatic radars (Sec. 1.3).

$$V_R = V \cos \alpha \qquad (1.2)$$

where α is the angle between the target velocity vector and the radar line-of-sight (LOS), (Fig. 1.1). It is found from the Doppler-frequency shift of the received signal, f_D, by

$$V_R = \frac{f_D c}{2f} = \frac{f_D \lambda}{2} \qquad (1.3)$$

where f is the radar frequency and λ is the wavelength.
- Target direction, found from the antenna beam orientation for maximum signal return.
- Target characteristics, found from the magnitude and features of the signal return, such as fluctuation characteristics, duration, and spectrum features.

1.2   Radar Functions

In most applications, radars perform one or more of the following functions.

- Search (also called surveillance), the examination of a volume of space for potential targets (Sec. 3.4).
- Detection, determining that a target is present (Sec. 3.3).
- Position measurement of target range, angular coordinates, and sometimes radial velocity (Sec. 3.5).

- Tracking, processing successive measurements to estimate target path (Sec, 3.6).
- Imaging, generating a two (or three) dimensional image of a target or area, frequently using synthetic-aperture processing (Sec. 5.4).
- Classification, discrimination and identification, determining the characteristics, type, and identity of a target (Sec. 5.5).

While many radars perform two or more functions in their normal operating modes, the term multi-function radar is applied to radars that are effective in a wide variety of functions, and usually can handle multiple targets. These radars usually have the following characteristics:

- Phased array antennas, for rapidly directing beams to the desired angular positions (Sec. 2.1).
- Multiple waveform types, for performing the various functions (Sec. 5.1).
- Digital signal processing, to handle the variety of waveforms and perform the desired functions on multiple targets (Sec. 2.5).
- Computer control, to schedule transmission and signal reception responsive to the information needs and the target configuration (Sec. 2.5).

1.3   Types of Radars

Monostatic radars have the transmit and receive antennas at the same location. Their separation is small compared with the target range. This allows both antennas to view the same volume of space. In many cases a single antenna is switched between transmit and receive functions. The radar can be located at a single site allowing simple transmit-receive coordination. The round-trip signal transmission time provides a direct measure of target range (Eq. 1.1), and the Doppler-frequency shift gives a measure of target radial velocity (Eq. 1.3).

Bistatic radars have transmit and receive antennas that are widely separated. This avoids interference between the transmitter and receiver and allows multiple passive receivers to operate with a single transmitter. The target reflection for some bistatic geometries may be advantageous (Sec. 3.1). The signal path consists of the range between the transmit antenna and the target, R_T, and the range between the target and the receive antenna, R_R. This sum defines an ellipsoid with the transmit and receive antennas as the foci (Fig. 1.2). It can be found from the time

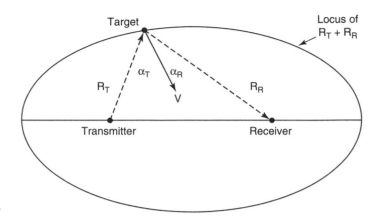

Fig. 1.2 Bistatic radar target-measurement geometry

between signal transmission and reception:

$$R_T + R_R = ct \tag{1.4}$$

The Doppler frequency shift is given by:

$$f_D = \frac{V(\cos \alpha_T + \cos \alpha_R)}{\lambda} \tag{1.5}$$

where α_T is the angle between the target velocity vector and the transmit LOS, and α_R is he angle between the target velocity vector and the receive LOS (Fig. 1.2).

Pulsed radars transmit a pulse and then listen for the reflected pulse return. This avoids interference between the transmitter and receiver, and enables use of a single antenna for both transmit and receive. The time from pulse transmission to pulse reception can readily be measured, allowing the target range to be found (for monostatic radars). Since monostatic radars can not receive until the pulse transmission is finished, the minimum range, R_M, of these radars is limited to:

$$R_M = \frac{c\tau}{2} \tag{1.6}$$

where τ is the pulse duration. When pulses are transmitted at a fixed pulse repetition interval (PRI, which is the reciprocal of the pulse repetition frequency, PRF), signal returns arriving after the next pulse has been transmitted may be interpreted as a return from a later pulse, giving an erroneous range. This will occur when the target range is greater than $nc(\mathrm{PRI}/2)$, and is called a second-time around-return when $n = 1$, and a multiple-time around-return when $n > 1$.

Continuous-wave (CW) radars transmit a continuous signal and receive simultaneously. The target radial velocity can be directly measured from the Doppler-frequency shift (for monostatic radars). If the frequency of the continuous wave is changed (frequency-modulated continuous wave, FMCW), the target range can also be measured. Interference between the transmitter and receiver may limit the power, and hence the sensitivity, of monostatic CW radars.

Coherent radars transmit waveforms derived from a stable frequency source (often called the STALO), which is also used in processing the received signals. This allows measurement of target radial velocity (Sec. 1.1), coherent pulse integration (Sec. 3.2), and rejection of background clutter by moving-target indication (MTI), pulse Doppler, and space-time adaptive processing (STAP) techniques (Sec 5.2 and 5.3). Noncoherent radars lack these capabilities.

Over-the-horizon (OTH) radars utilize ionospheric reflection to direct radar waves well beyond the normal radar horizon (Sec. 4.4). They operate in the high-frequency (HF) band (3–30 MHz), (Sec. 1.4). The frequency is selected to accommodate ionospheric conditions and target range. Very large antennas, high-power transmitters, and long processing times are required. The range and angle measurement accuracy is relatively poor, but may be suitable for warning.

Secondary-surveillance radars (SSR) transmit pulses from a rotating antenna with sufficient power to reach a cooperating aircraft. These aircraft employ a transponder that replies to the ground station at a different frequency. The responses contain coded pulses that supply information such as target altitude and identification. This technique provides long-range operation with the low power needed for one-way transmission, and avoids radar clutter return by using different up- and down-link frequencies. The military system, which uses classified response codes, is called identification friend or foe (IFF), (Sec. 5.5).

Synthetic aperture radars (SAR) transmit a series of pulses as the radar platform (aircraft or satellite) moves along its flight path. The return signals are processed to produce the effect of a very large aperture having very small beamwidth. This, along with good range resolution, allows generation of two-dimensional images of terrain and ground targets (Sec. 5.4).

Radar basing. The coverage and capabilities of radars is influenced by their basing:

- Terrestrial basing supports large, high-power radars, but they are line-of sight limited by the terrain horizon (Sec. 4.4).
- Sea basing supports large radars and allows radar relocation consistent with ship size and speed.
- Airborne radars can provide increased horizon range on low-altitude aircraft and ground targets, and can be relocated rapidly. However the aircraft payload may limit antenna size and transmitter power.
- Space-based radars can access any area of the earth, but their coverage may be constrained by satellite orbital characteristics. These radars operate at long ranges, and may be limited in antenna size and transmitter power.

1.4 Frequency Bands

Wavelength of electromagnetic radiation, λ, is related to its frequency, f, by:

$$\lambda = \frac{c}{f} \qquad f = \frac{c}{\lambda} \tag{1.7}$$

Radar Bands. The radar spectrum is divided into bands, most of which are designated by letters (Table 1.1). Within these bands specific frequency ranges are authorized for radar use by the International Telecommunications Union (ITU). Radars usually operate in a frequency range of about 10% of their center frequency. The operating frequency range is constrained both by the ITU allocations, and by bandwidth limitations of radar components.

Search radars often operate in the VHF, UHF and L bands because the large antenna sizes and high power needed are more easily obtained at these frequencies (Sec. 3.4). Tracking radars often operate in the X, K_U, and K bands because narrow beamwidth and good measurement precision are more easily obtained at these frequencies (Sec. 3.5). Multi-function surface radars often operate at the intermediate S and C band frequencies, and airborne fighter multi-function radars usually operate at X band.

Signal attenuation due to atmospheric absorption and rain increase with frequency (Sec 4.1 and 4.2), and can limit radar performance in the K_U, K, and K_A bands. The ionosphere can distort signals passing

TABLE 1.1 Radar Frequency Bands [8]

Band Designation	Frequency Range	Assigned Radar Frequency Ranges	Common Radar Frequency	Common Radar Wavelength
HF	3–30 MHz			
VHF	30–300 MHz	138–144 MHz	220 MHz	1.36 m
		216–225 MHz		
UHF	300–1,000 MHz	420–450 MHz	425 MHz	0.71 m
		890–942 MHz		
L	1–2 GHz	1.215–1.4 GHz	1.3 GHz	23 cm
S	2–4 GHz	2.3–2.5 GHz	3.3 GHz	9.1 cm
		2.7–3.7 GHz		
C	4–8 GHz	4.2–4.4 GHz	5.5 GHz	5.5 cm
		5.25–5.925 GHz		
X	8–12 GHz	8.5–10.68 GHz	9.5 GHz	3.2 cm
Ku	12–18 GHz	13.4–14 GHz	16 GHz	1.9 cm
		15.7–17.7 GHz		
K	18–27 GHz	24.05–24.25 GHz	24.2 GHz	1.2 cm
		24.65–24.75 GHz		
Ka	27–40 GHz	33.4–36 GHz	35 GHz	0.86 cm
V	40–75 GHz	59–64 GHz		
W	75–110 GHz	76–81 GHz		
		92–100 GHz		
Millimeter wave	110–300 GHz	126–142 GHz		
		144–149 GHz		
		231–235 GHz		
		238–248 GHz		

through it at VHF and UHF frequencies (Sec. 4.6). Ionospheric reflection is utilized by OTH radars (Sec. 1.3).

ITU Bands. The ITU designation for bands within the radar spectrum is given in Table 1.2.

Electronic Warfare Bands. The letter band designations used by the electronic warfare (EW) community are given in Table 1.3.

1.5 Military Nomenclature

United States military electronic systems are specified using the Joint Electronics Type Designation System (JETDS), formally called the Joint Army-Navy Nomenclature System (AN System). The format is:

AN/ABC - # D suffix

TABLE 1.2   International Telecommunications Union Frequency
Bands [8]

Frequency Range	Band Designation	Metric Designation
3–30 MHz	High Frequency (HF)	Dekametric Waves
30–300 MHz	Very-High Frequency (VHF)	Metric Waves
0.3–3 GHz	Ultra-High Frequency (UHF)	Decimetric Waves
3–30 GHz	Super-High Frequency (SHF)	Centimetric Waves
30–300 GHz	Extremely-High Frequency (EHF)	Millimetric Waves

TABLE 1.3   Electronic Warfare Frequency
Bands [9]

EW Band	Frequency Range
A	30–250 MHz
B	250–500 MHz
C	500–1,000 MHz
D	1–2 GHZ
E	2–3 GHz
F	3–4 GHz
G	4–6 GHz
H	6–8 GHz
I	8–10 GHz
J	10–20 GHz
K	20–40 GHz
L	40–60 GHz
M	60–100 GHz

Where:

- AN indicates that the JETDS system is being used.
- The first letter in the 3-letter group following the slash, (A here),
  specifies the equipment installation.
- The second letter in the group, (B here), specifies the type of
  equipment.
- The third letter in the group, (C here), specifies the purpose of the
  equipment.

TABLE 1.4   Letter Codes for JETDS, (AN), System

Letter	Code	Meaning
Installation	A	Piloted aircraft
(first letter)	D	Pilotless carrier (drone, UAV)
	F	Fixed ground
	G	General ground use
	M	Ground mobile
	P	Human portable
	S	Water (surface ship)
	T	Transportable (ground)
	V	Vehicle (ground)
Type	L	Countermeasures
(second letter)	M	Meteorological
	P	Radar
	S	Special or combination
Purpose	G	Fire control
(third letter)	N	Navigation aid
	Q	Special or combination
	R	Receiving or passive detecting
	S	Detection, range and bearing, search
	X	Identification or recognition
	Y	Surveillance and control

- The number following the dash, (# here), specifies the model number of this type of equipment.
- The next letter, (D here), if present, specifies the equipment modification.
- A suffix (V) followed by a number, if present, indicates an equipment version number.
- A suffix (T), if present, indicates a training system.

The letter codes for the three-letter group that are relevant to radar systems are given in Table 1.4.

1.6 Radar Configurations

A typical configuration is shown in Fig. 1.3 for a pulsed monostatic radar using a single antenna for transmit and receive and operating coherently. The antenna is switched between the transmitter and receiver by a

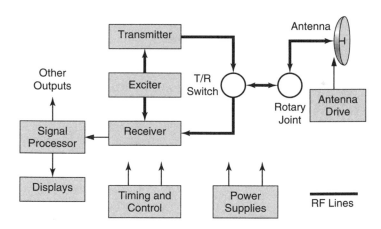

Fig. 1.3   Typical radar configuration

transmit/receive (T/R) switching device. Many of the features and components of this configuration are used in the other radar types described (Sec. 1.3).

- The exciter includes a stable frequency source (STALO), and a waveform generator. Its output is fed to the transmitter, and to the receiver as a frequency reference for coherent processing. Noncoherent radars may not employ an exciter; then the waveform is generated by an oscillator in the transmitter (Sec. 2.2).
- The transmitter amplifies the signal from the exciter to provide the radio-frequency (RF) power level to be transmitted. The transmitter may also contain a modulator to turn the transmitter on during the pulse transmissions (Sec. 2.2).
- The transmitter output is connected to the antenna using a high-power RF transmission medium. At microwave frequencies, this is usually waveguide. Co-axial cable or structures are often used at UHF and VHF. A similar transmission medium is used to connect the antenna to the receiver, but at lower power levels.
- The antenna radiates the transmitted RF power, and collects the reflected signals that impact it. The antenna usually provides a directive transmit and receive pattern to focus the radar observations in a direction of interest and allow measurement of target angle (Sec. 2.1).

- When antennas scan mechanically to observe the desired regions, a means is needed to connect the antenna to the transmitter and receiver. This is often provided by one or more waveguide rotary joints. Where the mechanical scanning angles are limited, flexible cables may be used instead. Antenna drive motors and their associated control mechanism are also used with mechanically-scanning antennas.

- When a single antenna is used for both transmitting and receiving, a transmit-receive (T/R) switch (also called a duplexer), connects the antenna to the transmitter during pulse transmission, and to the receiver during the listening period. A microwave circulator is often used for this function. When separate antennas are used for transmit and receive, a protection device is usually used to prevent the high transmit power levels from damaging the receiver. Such a device is often used along with the circulator to provide further protection when a single antenna is used.

- The receiver amplifies the received signals and converts them to lower intermediate-frequency (IF), or video signals. In most modern radars, these signals are digitized and processed in the signal processor. In some cases, analog processing in the receiver is used for detection and measurement (Sec. 2.3).

- The signal processor filters the received signals, performs detection processing, generates measurements of target position and velocity, tracks targets, and extracts other target features of interest. The resulting information is used in radar displays and transmitted to other users (Sec. 2.5).

- Some radars employ a single waveform and simple timing that can be generated in the exciter. More complex radars, especially multi-function radars, employ a timing and control system that selects and schedules transmission of waveforms, as well as receiver listening periods and processing modes.

- Power supplies convert prime power to provide the voltage levels and stability required by the various radar components.

2 | Radar Subsystems

2.1 Antennas

Antenna parameters. Directivity, D, is the maximum radiation intensity divided by the average radiation intensity.

Gain, G, is the maximum radiation intensity divided by the radiation intensity from a lossless isotropic source. It differs from directivity by including antenna loss factors. These are:

Antenna ohmic loss, L_O.

Antenna efficiency loss, L_E.

$$G = \frac{D}{L_O L_E} \tag{2.1}$$

Effective aperture area, A, determines the signal power collected by the antenna. It is related to antenna gain by (Fig. 2.1):

$$A = \frac{G \lambda^2}{4 \pi} \qquad G = \frac{4 \pi A}{\lambda^2} \tag{2.2}$$

The effective aperture area is usually smaller than the physical antenna area, A_A, due to the antenna losses:

$$A = \frac{A_A}{L_O L_E} \qquad G = \frac{4 \pi A_A}{\lambda^2 L_O L_E} \tag{2.3}$$

The antenna beamwidth, θ, is usually defined at the level of half the power of the beam peak, and is referred to as the 3 dB beamwidth (Fig. 2.2):

$$\theta = \frac{k_A \lambda}{W} \tag{2.4}$$

where k_A is the antenna beamwidth coefficient, usually near unity, and W is the antenna dimension in the plane of the beamwidth. The beamwidth can be defined in two orthogonal planes. These are often the x and y

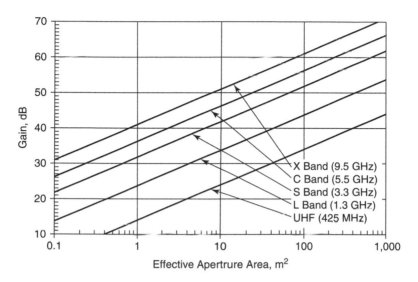

Fig. 2.1 Antenna gain vs. effective aperture area for four frequencies

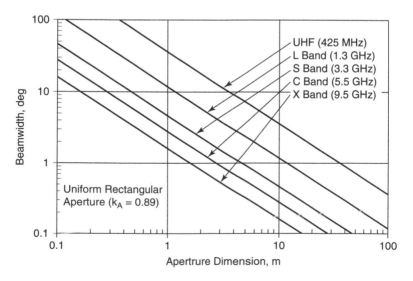

Fig. 2.2 Antenna beamwidth vs. aperture width for four frequencies

planes, (θ_X, θ_Y), where the y dimension is vertical. For antennas with a horizontal beam direction, these correspond to the azimuth and elevation coordinates (θ_A, θ_E).

The antenna gain can be estimated from these orthogonal beamwidths [10, App. A]:

$$G \approx \frac{10.75}{\theta_X \theta_Y L_O} \quad G \approx \frac{10.75}{\theta_A \theta_E L_O} \tag{2.5}$$

When separate antennas are used for transmit and receive, or when antenna losses are different in the transmit and receive modes, gain, effective aperture area, and beamwidth can be specified for each mode: $G_T, G_R, A_T, A_R, \theta_T, \theta_R$.

Antenna sidelobes can be specified relative to the main-beam, SL, (a number less than unity or a negative dB value), or relative to a lossless isotropic source, SLI. These are related by:

$$\text{SLI} = \text{SL}\, G \tag{2.6}$$

When these are expressed in decibels, dB, (Sec. 6.3):

$$\text{SLI(dB)} = \text{SL(dB)} + G\text{(dB)} \tag{2.7}$$

The antenna pattern is formed in the far-field of the antenna, also called the Fraunhofer region. The range, R_F, at which this begins is given by:

$$R_F = \frac{2\,W^2}{\lambda} \tag{2.8}$$

Antenna pattern. The antenna pattern in the far field is given by the Fourier transform of the aperture illumination function:

$$G(\psi) = \left[\int a(x) \exp\left(-j\,2\pi\frac{x}{\lambda}\sin(\psi)\right) dx \right]^2 \tag{2.9}$$

where ψ is the angle from the antenna mainbeam center, and $a(x)$ is the antenna current density as a function of the distance from the antenna center, also called the aperture illumination.

Weighting (also called tapering) of the aperture illumination function can reduce the close-in sidelobes, but it increases the beamwidth and the aperture efficiency loss. (With weighting in two orthogonal antenna dimensions, the aperture efficiency loss is the product of the losses for the two weighting functions used.) The characteristics of several common weighting functions are given in Table 2.1. The cosine^x, truncated Gaussian, and cosine on a pedestal functions are easily implemented by

TABLE 2.1   Antenna Characteristics for Aperture Weighting Functions
(data from [10, App. A])

Aperture Weighting	Aperture Efficiency	Aperture Efficiency Loss, dB	Antenna Beamwidth Coefficient, k_A	First Sidelobe Level, dB
Uniform rectangular	1.0	0	0.886	−13.3
Uniform circular	1.0	0	1.028	−17.6
Cos	0.802	0.956	1.189	−23.0
Cos^2	0.660	1.804	1.441	−31.5
Cos^3	0.571	2.44	1.659	−39
Cos^4	0.509	2.93	1.849	−47
Cos^5	0.463	3.34	2.03	−54
Truncated Gaussian 50% edge illumination	0.990	0.0445	0.920	−15.5
Truncated Gaussian 14% edge illumination	0.930	0.313	1.025	−20.8
Truncated Gaussian 1.9% edge illumination	0.808	0.928	1.167	−32.1
Truncated Gaussian 0.5% edge illumination	0.727	1.387	1.296	−37
Cos on 50% pedestal	0.965	0.1569	0.996	−17.8
Cos on 30% pedestal	0.920	0.363	1.028	−20.5
Cos on 20% pedestal	0.888	0.516	1.069	−21.8
Cos on 10% pedestal	0.849	0.711	1.121	−22.9
Cos on 5% pedestal	0.827	0.827	1.151	−23.1
Taylor 20 dB sidelobes	0.951	0.218	0.983	−20.9
Taylor 25 dB sidelobes	0.900	0.455	1.049	−25.9
Taylor 30 dB sidelobes	0.850	0.704	1.115	−30.9
Taylor 35 dB sidelobes	0.804	0.948	1.179	−35.9
Taylor 40 dB sidelobes	0.763	1.178	1.250	−40.9

feed horns of reflector antennas and produce sidelobes that decrease with
distance from the main beam. Taylor weighting is more readily implemented in array antennas, and produces nearly uniform close-in sidelobe
levels and relatively-low aperture efficiency losses.

The weighting functions produce very low sidelobes at angles far
from the main beam, but larger far-out sidelobes are generated by:

- Errors in the antenna shape or illumination pattern.
- Reflections from antenna structural elements.
- Spillover from reflector illumination.

Many radars employ monopulse feeds to measure target angle with a
single pulse (Sec. 3.5). These generate two beams that are slightly offset

in angle from the main beam with their signals subtracted to produce a difference pattern. Two such beam pairs are used to measure angle in two orthogonal coordinates. In reflector antennas, these beams are produced by a cluster of four or more feed horns.

Signal polarization. The polarization of the radiated signal is defined as the orientation of the electric-field vector. Linear polarization, horizontal (H) or vertical (V) is often used. In circular polarization, the electric-field vector rotates at the signal frequency, clockwise for right circular polarization (RC) or counter-clockwise for left circular (LC) polarization.

Most antennas transmit and receive a single polarization. (For circular polarization, the polarization received is usually the opposite rotation direction from that transmitted). Antennas can be designed to transmit and receive two orthogonal polarizations, e. g. H and V, or RC and LC. Only one polarization is transmitted at a time, but antennas can be switched between polarizations for successive transmit pulses. Two orthogonal polarizations can be received simultaneously, and processed in separate receive channels (Sec. 2.3).

Reflector antennas employ solid or mesh metallic reflectors, illuminated by feed horns, and shaped to produce the desired far-field antenna pattern. Common reflector antennas are:

- Dish antenna, a circular reflector with a parabolic shape that is steered mechanically in two angular coordinates. This generates a narrow symmetrical beam, often called a pencil beam. Dish antennas are well suited for observing and tracking individual targets.
- Rotating reflector antennas have a parabolic contour in the horizontal plane, generating a narrow azimuth beam. The vertical contour and feed are shaped to generate the desired elevation-beam coverage. These are often called fan beams, and usually rotate continuously in azimuth to provide 360-degree search coverage with periodic target updates.

Planar-array antennas employ an array of elements radiating in phase and produce a beam shape that depends on the array shape and weighting. They are rotated or steered mechanically to direct the beam. Common implementations are:

- Dipole radiating elements arrayed over a plane reflector. These are usually used at VHF and UHF.

- Slotted waveguide arrays that employ an array of waveguides with slots in each waveguide to provide the radiating elements. These are usually used at microwave frequencies.

Phased-array antennas also employ an array of radiating elements, usually in a planar configuration. However the phase of each element is individually controlled using electronically-controlled phase shifters, enabling the beam to be rapidly steered in the desired direction. This supports interlacing of search, track, and other functions to meet tactical requirements. Many phased array radars have repertoire of waveforms and support multi-function operation.

When a phased array has n_E identical elements, each with gain G_E and effective aperture area, A_E, the array gain and effective aperture area are the sum of the element values:

$$G = n_E G_E \qquad A = n_E A_E \qquad (2.10)$$

This corresponds to uniform aperture illumination. When aperture weighting on receive is desired, signals from the appropriate elements are attenuated. When weighting is desired on transmit, power levels are reduced for the appropriate elements. In either case the effect is characterized by an aperture efficiency loss (Sec. 2.4).

When the antenna beam is scanned off the array broadside angle at a scan angle φ:

- The gain, G_φ, and effective aperture area, A_φ, are reduced due to the reduced projected array area in the beam direction:

$$G_\varphi = G \cos \varphi \qquad A_\varphi = A \cos \varphi \qquad (2.11)$$

- The gain and effective aperture area are further reduced by the gain and effective aperture area of the array element at that scan angle. These two effects are incorporated in a two-way scan loss, L_S.
- The beamwidth, θ_φ, is broadened by the reciprocal of the cosine of the scan angle:

$$\theta_\varphi = \frac{\theta}{\cos \varphi} \qquad (2.12)$$

Transmit signals can be distributed to the array elements and received signals from the array elements to the receiver by:

- Space feed. The signals are radiated and received by feed horns that illuminate the array face. Phase shifters at each element adjust the transmit and receive signal phases appropriately.

- Waveguide. The signals are distributed to and from the elements by waveguide, or sometimes coaxial cable. Phase shifters at each element adjust the signal phase.
- Transmit/Receive (T/R) modules (Sec. 2.4). A module at each element provides the final stage of transmit amplification, the first stage of low-noise reception, and phase shifting of the transmitted and received signals. The signals are distributed to and from the T/R modules by one of the above techniques.

Types of phased arrays include [5, Ch. 3]:

- Full-field-of-view (FFOV) phased arrays can scan electronically to about 60 degrees off array broadside. The array elements are spaced at about 0.6 λ, and have gain of about 5 dB. The array has a scan loss of about:

$$L_S \approx \cos^{-2.5} \varphi \qquad (2.13)$$

Three or more FFOV arrays are needed to provide hemispheric radar coverage.
- Thinned FFOV arrays employ dummy or missing elements in some array positions. This results in fewer active array elements, lower gain and effective aperture area, and higher sidelobe level than for filled arrays of similar size.
- Limited-field-of-view (LFOV) phased arrays are designed to scan off-broadside to a maximum scan angle φ_M that is less than 60 degrees. They require fewer, but higher-gain elements than FFOV arrays. The increased element spacing, d, is limited by the requirement to avoid unwanted antenna lobes called grating lobes:

$$d \leq \frac{0.5 \lambda}{\sin \varphi_M} \qquad (2.14)$$

When a phased array is scanned off broadside by an angle φ, the difference in signal arrival time across the array limits the signal bandwidth, B, to:

$$B \leq \frac{c}{W \sin \varphi} \qquad (2.15)$$

where W is the array dimension in the scan plane. Using time-delay steering, rather than phase steering eliminates this limitation. Time-delay steering need only be applied to subarrays having dimension, W_S, that satisfy Eq. 2.15:

$$W_S \leq \frac{c}{B \sin \varphi} \tag{2.16}$$

These limitations apply to both orthogonal scan directions.

Hybrid antennas can be configured to combine features of reflectors (simplicity and low cost), with the flexibility of phased arrays. Typical configurations include:

- Rotating reflector or array antennas that generate stacked beams in elevation angle
- Phased arrays that are mechanically steered to position their field-of-view in the desired direction.

2.2 Transmitters

Transmitter parameters. Average transmitter RF power output, P_A.

Peak transmitter RF power output, P_P, which is larger than the average power in pulsed radars. These are related by the transmitter duty cycle, DC:

$$\text{DC} = \frac{P_A}{P_P} \tag{2.17}$$

The transmitted pulse duration, τ, in pulsed radars. Many radars, especially phased arrays, have a repertoire of pulses with various durations. When a radar operates at a fixed pulse repetition frequency, PRF, the duty cycle is:

$$\text{DC} = \tau \text{PRF} \tag{2.18}$$

The pulse energy, E, is given by:

$$E = P_P \tau \tag{2.19}$$

The maximum transmitter pulse duration, $\tau_{\max}$, may be limited by the heat build-up in transmitter components, especially vacuum tubes. The maximum pulse energy is then:

$$E_{\max} = P_P \tau_{\max} \tag{2.20}$$

The transmitter efficiency, η_T, is given by:

$$\eta_T = \frac{P_A}{P_S} \tag{2.21}$$

where P_S is the prime power supplied to the transmitter. Transmitter efficiency is typically 15–35%. Note that this value normally will be lower than the efficiency of the RF output device due to losses and power used by other transmitter components.

The overall radar efficiency is lower than η_T, due both to waveguide and antenna losses, and to power use by other radar components. It is typically in the 5–15% range

The transmitter center frequency, f, and the signal bandwidth available for signal transmission, B_T, are key parameters and can significantly influence the transmitter design. The stability of the transmitter frequency is also important for coherent radar functions. The frequency must remain stable with low phase noise for the coherent-processing duration.

Transmitter configuration. A simplified diagram of a coherent transmitter showing exciter and power-amplifier components is shown in Fig. 2.3.

The radar transmitter frequency is generated by a stable local oscillator (STALO), which also provides a reference signal for processing the received signals. For a radar capable of transmitting at several frequencies, the STALO employs a set of stable oscillators, which are usually phase locked to an extremely stable crystal oscillator. For noncoherent radars, the stability of the local oscillator is not important, and in some cases the transmitter itself operates as an oscillator.

The waveform generator produces the signal modulation of the transmitted waveform (Sec. 5.1). For simple pulses, this simply requires gating the RF signal. More complex waveforms also require precise phase or frequency modulation. These can be implemented using analog devices such as a bi-phase modulator or surface-acoustic wave (SAW) delay line. Digital waveform generation is used in many modern multi-function radars

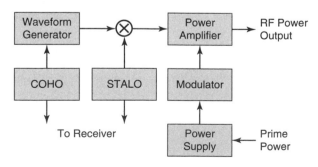

Fig. 2.3   Coherent transmitter configuration

(Sec. 2.5). Many radars employ a coherent local oscillator (COHO) to provide a stable signal for the waveform generator and a reference for signal demodulation in the receiver.

The STALO and waveform signals are combined by a mixer, and fed to the transmitter power amplifier. This may have several stages to increase the power to the desired level. In some cases the power from two or more parallel output devices may be combined to provide the desired power. Note that the transmit power, P_P and P_A, are specified at the transmitter output. Any losses within the transmitter for combining or directing the power will affect these values.

The prime power is converted to the direct-current (DC) voltages required by the transmitter. Pulsed radars usually employ a modulator to provide the direct-current power to the transmitter during the pulse periods. This may simply be an electronic switch, but in many cases modulators also employ energy storage devices such as capacitors, pulse-forming networks, and delay lines.

The transmitter also requires cooling devices to remove the heat generated. Since transmitter efficiency is rarely greater than 35%, the heat generated is usually more than twice the transmitter RF power output.

RF power sources. The selection of the transmitter RF power source depends on many factors, including output power level, frequency, bandwidth, efficiency, gain, stability and noise. Power sources that function as oscillators are used in many noncoherent radars, while those that function as amplifiers are usually preferred for coherent radars. Key characteristics of common RF power sources are summarized in Table 2.2, and discussed below [11, 12].

- Magnetrons are crossed-field microwave vacuum tubes that have perpendicular electric and magnetic fields. They operate as pulsed oscillators, generating simple fixed-frequency pulses at high power

TABLE 2.2   Characteristics of RF Power Sources (data from [11])

RF Power Source	Frequency Range, GHz	Bandwidth, %	Peak Pulse Power, MW	Duty Cycle, %
Magnetron oscillator	1–90	0.1–2	10	1–10
Crossed-field amplifier	1–30	10–20	5	1–10
Triode	0.1–1	2–10	5	1–10
Klystron	0.1–300	5–10	10	1–10
Traveling-wave tube (coupled cavity)	1–200	10–20	0.25	1–10
Solid-state amplifier	0.1–20	20–40	0.01–1	20–50

levels. While their signal is relatively narrowband, they can be rapidly mechanically tuned over a 1 to 15 percent bandwidth. They are normally used in noncoherent radars, although injection locking can provide a degree of coherence.

- Crossed-field amplifiers also have perpendicular electric and magnetic fields, but function as microwave amplifiers with high power levels and modest gains, typically 15 dB. They are used in coherent radars and can provide wide signal bandwidth, but have relatively high phase noise.

- Triode vacuum tubes are used as power amplifiers at VHF and UHF. They have good phase stability and high peak power levels.

- Klystrons are linear-beam vacuum tubes that have parallel electric and magnetic fields aligned along the tube axis. Interaction between the electron beam and the RF field occurs along the length of the tube. They can operate over the entire radar spectrum, and are capable of high peak and average power. They have excellent phase stability and gains of 40–60 dB.

- Traveling-wave tubes (TWTs) are also linear-beam vacuum tubes. They have high gain, wide bandwidth and excellent phase stability. The coupled-cavity configuration provides modest power levels (250 kW peak), while the helix configuration provides lower output power (20 kW peak), and supports a signal bandwidth of 2-3 octaves (factor of 4–8). They are often used in driver stages of transmitters that require higher output power levels.

- Solid-state amplifiers, such as gallium nitride (GaN) high electron mobility transistors (HEMT), offer the advantages of low-voltage operation, signal bandwidth as high as 50%, high duty cycle, and excellent reliability. The power of individual devices is limited to 10–1,000 W, with the power capability decreasing with increasing frequency. Multiple solid-state devices can be used to increase transmitter power. Solid state devices are often used in transmit/receive modules employed by phased arrays (Sec. 2.4).

2.3   Receivers

Receiver parameters. The receiver noise temperature, T_R, characterizes the noise produced by the receiver, measured at the receiver input. This is one component of the system noise temperature, T_S, used to calculate the signal-to-noise ratio (Sec. 3.2).

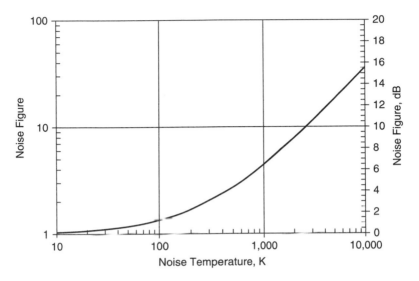

Fig. 2.4   Noise figure vs. noise temperature

The receiver noise power, P_R, is related to T_R by:

$$P_R = kT_R B_R \qquad T_R = \frac{P_R}{k\,B_R} \qquad (2.22)$$

where k is Boltzmann's constant and B_R is the receiver bandwidth. The value of $k = 1.38 \times 10^{-23}$ joules per Kelvin (Sec. 6.2).

The receiver noise figure (sometimes called noise factor), F_R, is related to T_R, by (Fig. 2.4):

$$F_R = \frac{T_R}{290} + 1 \qquad T_R = 290(F_R - 1) \qquad (2.23)$$

The receiver noise level is usually dominated by noise from the first stage of the receiver, since it is amplified by that stage before noise from later stages is added. For this reason, the first stage is usually a low-noise amplifier (LNA).

The effect on T_R of noise from later stages can be found from:

$$T_R = T_1 + \frac{T_2}{G_1} + \frac{T_3}{G_1 G_2} + \cdots \qquad (2.24)$$

where T_1, T_2, and T_3, are the noise temperatures of the first, second and third receiver stages and G_1 and G_2 are the gains of the first and second stages.

When using noise figures:

$$F = F_1 + \frac{F_2 - 1}{G_1} + \frac{F_3 - 1}{G_1 G_2} + \cdots \quad (2.25)$$

where F_1, F_2, and F_3 are the noise figures for the stages.

The receiver dynamic range is the range of signal levels for which the receiver provides linear amplification. It is usually defined as the range between the noise floor and the level at which the signal is compressed by 1 dB [13]. Signals that exceed the dynamic range interfere with the separation of signals and interference. Receivers employ various forms of gain control to maintain the signals within their dynamic range:

- Manual gain control by an operator
- Automatic gain control (AGG), which reduces the gain to avoid saturation
- Sensitivity-time control (STC), which reduces the gain at short ranges to avoid large returns from targets at these ranges.

Receiver configuration. A typical superheterodyne receiver, the most common receiver configuration, is shown in Fig. 2.5. The incoming signal is first amplified in the LNA, and then mixed with the STALO signal from the exciter to produce an intermediate-frequency (IF) signal. In the IF amplifier, the signal is filtered to accept the signal band but reject noise and spurious signals outside that band.

In the configuration shown, the IF signal is synchronously detected, using the COHO from the exciter, to produce in-phase, (I), and quadrature, (Q), video signals. These are digitized and fed to the signal processor for further processing. The number of bits used in the digitization is in the 6–14 range, depending on the quantization level and the dynamic

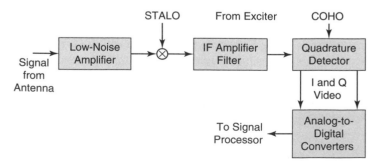

Fig. 2.5   Receiver channel configuration

range. The sampling rate for both I and Q signals must exceed the signal bandwidth. Equivalently, the signal amplitude and phase can be detected rather than the I and Q signals.

In other receiver configurations, analog IF signals feed the signal processor for analog processing (Sec. 2.5). Alternatively, signals may be digitized earlier in the receiver, at IF or even RF. In many noncoherent receivers, the IF amplifier output is simply amplitude detected.

While some radars operate with a single receiver channel, most have multiple channels to perform their various functions, including:

- Sum channel for target detection and reference for monopulse measurement
- Monopulse difference channels for target angular measurements. Two channels are usually provided for azimuth and elevation angle measurement (Sec 2.1).
- Auxiliary reference channels for sidelobe blankers or sidelobe cancellers used to reject interference (Sec. 4.7).
- Channels for dual-polarization reception (Sec. 2.1).

Low-noise amplifiers, (LNA). Types of low-noise amplifiers are discussed below [14]:

- Transistor amplifiers are used in most modern radars because of their simplicity and low-noise performance. Gallium arsenide field-effect transistor, (GaAsFET), amplifiers and high electron mobility transistors, (HEMT), can provide noise temperatures of 100–300 K and gains of about 30 dB throughout the microwave frequency range.
- Tunnel-diode amplifiers are negative-resistance devices that provide modest gains and noise temperatures of about 500 K.
- Parametric amplifiers, (also negative resistance devices), and masers provide noise temperatures of 100-300 K when not cooled. When cooled to liquid nitrogen temperature, (77 K), noise temperatures of less than 50 K can be obtained. However, these are large, complex devices that are used in radio astronomy and similar applications, and rarely used for radar.
- Traveling-wave tube (TWT) amplifiers can provide very-wide signal bandwidth (1–2 octaves), and high gain (40 dB), with noise temperatures in the 300 K range.
- Vacuum tubes are rarely used for LNAs because their noise temperatures are of the order of 1,000 K.

2.4   Transmit/Receive Modules

A popular method for implementing phased-array radars (Sec. 2.1), employs transmit/receive, (T/R), modules. These are usually solid-state devices that contain a transmitter, a low-noise amplifier, (LNA), an adjustable phase shifter and attenuator, and switches to control the signal path during transmit and receive periods.

The module is connected directly to an antenna radiating element. This results in very-low microwave losses, and eliminates losses from phase shifters following the transmitter stage and preceding the LNA. It also allows multiple low-power transmitter amplifiers to be used, without combining networks, to produce high transmit power levels.

In most radar configurations, a T/R module feeds each array element, but in some cases a module may feed two or more elements, or two or more modules may feed an array element.

The total transmitted peak and average power are:

$$P_P = n_M P_{PM} \qquad P_A = n_M P_{AM} \qquad (2.26)$$

where n_M is the number of modules, and P_{PM} and P_{AM} are the module peak and average powers. The receiver noise temperature T_R is equal to the module noise temperature T_{RM}.

T/R module configuration. A typical module configuration is shown in Fig. 2.6. During transmit periods, the signal from the exciter driver is fed to a variable attenuator and a variable phase shifter, both of which are controlled by digital commands from the beam-forming and steering computer to form the transmit beam. The signal is then amplified in the transmitter power amplifier and fed by the circulator to the array element.

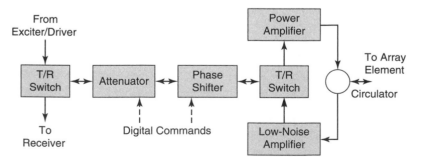

Fig. 2.6   Transmit/receive module configuration

The received signals from the array element are fed by the circulator to a low-noise amplifier. During receive periods these are then fed through the phase-shifter and attenuator, which may be reset from their transmit values, to produce the desired receive beam, and combined in the receiver.

T/R module components. The major components of a solid-state T/R module are described below [15]:

- The transmitter is usually a solid-state amplifier with power output consistent with transistor capabilities. Peak power is typically in the 10–1,000 W range, depending on radar frequency (Sec. 2.2).
- The receiver generally employs a gallium arsenide field-effect transistor, (GaAsFET), or high electron mobility transistor, (HEMT), transistor, providing noise temperature in the 100–300 K range (Sec. 2.3).
- A circulator is usually employed to direct the transmitter signal to the array element and the received signal from the array element to the LNA. Additional devices may be used to provide further receiver protection.
- The phase shifter is usually a binary device that adjusts the signal phase in increments of 180, 90, 45, 22.5 degrees, for a typical 4-bit phase shifter. This is accomplished by switching delay-line segments with incremental lengths into and out of the signal path using diode switches.
- The signal attenuator is often also a binary device that switches the various increments of attenuation into and out of the signal path.
- Diode switches are used to route the transmit and receive signals through the attenuator and phase shifter as required during the transmit and receive interval.

The variable attenuator can be used to apply aperture weighting. Weighting is usually not used for transmit in order to maximize the transmitted power. It is often used in receive to reduce the close-in sidelobes (Sec. 2.1). Weighting applied this way can be treated as an aperture efficiency loss relative to the unweighted gain and effective aperture area (Eq. 2.10).

T/R modules are often packaged in a box having two dimensions consistent with the element spacing on the array face (about 0.6 λ for full-field-of-view arrays – see Sec 2.1). The modules can then fit directly behind the array face, and extend back a distance equal to the third module dimension. Because of its resemblance to a brick wall, this is called a brick configuration.

Technology is being developed to fabricate modules using several shallow layers for integrated circuits, interconnections, power supply, etc. Subarrays of multiple modules will be manufactured this way. This is called a tile configuration, because of the shallow depth of such modules. The tile concept is attractive for applying conformal arrays to curved surfaces such as aircraft exteriors.

2.5   Signal and Data Processing

Radar processing functions depend on the functions to be performed by the radar and the nature of the outputs to be generated. Simple search radars may only detect and display the amplitude of the IF signal. Radars with more complex functions require more complex processing, and multi-function radars often require configuring the processor to the function being performed.

Processing functions. Radar signal and data processing functions can include:

- Waveform generation. This ranges from simple constant-frequency pulses, to more complex pulse-compression waveforms such as linear frequency modulation and phase-reversal pulses (Sec 5.1). (While the waveform generator is part of the exciter (Sec 2.2), it employs the same techniques discussed here.)
- Matched filtering. The receiver band-pass often provides an adequate matched filter for constant-frequency pulses. Pulse-compression waveforms, such as those mentioned above, require further processing to achieve their inherent pulse-compression gain and range resolution (Sec 5.1).
- Clutter-reduction processing. Techniques for reducing interference from terrain and sea clutter include moving-target indication (MTI), (Sec. 5.2), pulse-Doppler filtering (Sec 5.3), and space-time adaptive processing (STAP), (Sec. 5.4). These involve coherent processing the returns from two or more successive pulses, and in the case of STAP, from individual array elements as well.
- Pulse integration. Summing the signal returns from successive pulses increases the signal-to-noise ratio (Sec. 3.2). It can be done noncoherently, using only the signal amplitudes, or coherently, using the phase information as well.
- Target detection. This is usually done by setting a threshold level and designating signals that exceed the threshold as potential

targets. The threshold is set to exclude most samples of background noise, and may be adjusted to maintain a desired false-alarm rate when faced with varying noise levels, e.g., from jammers, using a constant-false-alarm (CFAR) technique (Sec. 3.3).

- Target measurement. Target parameters measured usually include range, azimuth and elevation angles. Radial velocity may be measured either by successive range measurements or from Doppler-frequency shift (Sec 3.5). Other target characteristics, such as radar cross section, radial length, and Doppler-frequency spectrum, may be measured to support target classification (Sec. 5.5).

- Target tracking. Successive target measurements are combined using tracking filters to estimate the target path. Filter characteristics may be fixed, or vary depending on the nature of the target. Radars that track multiple targets require associating signal returns with the correct target track or establishing a new track (Sec. 3.6).

- Image generation. When a radar is moving past a target, or a target is moving past a radar, a two-dimensional target image may be generated by coherently combining signal returns from multiple pulses. Synthetic-aperture radar (SAR) and inverse synthetic-aperture radar (ISAR) have this capability, which is often used to generate radar maps of terrain (Sec. 5.4 and 5.5).

- Sidelobe blanking and canceling. These techniques mitigate effects of radar jammers. They employ auxiliary antennas and receiver channels to sense the jammer signal and either deactivate the receiver during the jammer pulse, or coherently subtract the jammer signal from the receiver channel (Sec. 4.7).

Analog processing techniques. Analog processing is used in many existing radars, and in new radars where the requirements are not demanding. Analog techniques include:

- Waveform generation uses signal gates for constant-frequency pulses, dispersive devices such as surface-acoustic wave (SAW) delay lines for linear FM waveforms, and switched delay lines for phase-coded waveforms (Sec. 5.1).

- Matched filtering employs band-pass filters, dispersive devices and switched delay lines.

- Delay lines and signal gates can be used to implement MTI and pulse integration.

- Filter banks provide Doppler-frequency measurement and pulse-Doppler processing.

- Analog feedback techniques implement CFAR (Sec. 3.3), and sidelobe-canceller processing, (Sec. 4.7).
- Closed-loop servo techniques enable target tracking.
- Optical techniques generate SAR images.

Digital processing techniques. Most analog techniques are limited in their stability, accuracy, and flexibility, especially in coherent multi-function radars. As a result, digital signal and data processing is widely used in modern radars.

Digital waveform generation is done in the exciter using a digital frequency synthesizer that switches waveform segments under control of a digital processor.

In coherent radars, the receiver IF amplifier output is often digitized into in-phase (I) and quadrature (Q) components, using a coherent reference. Digital processing is then used to perform the functions that are to be implemented in the radar. These functions can be grouped as follows [16].

- The initial stages of signal processing perform functions such as matched filtering; MTI, pulse-Doppler, or STAP; pulse integration; and threshold setting. These generally improve the signal-to-noise ratio, reduce interference, and produce potential targets for further processing. Such functions are performed on the incoming data, and can require very high computational rates, of the order of 10^9 to 10^{12} operations per second (OPS), and radars often employ custom processors to achieve these rates.
- The potential targets are then processed to extract desired data. This includes measurement of target position and other characteristics. The data rates here are more modest and are met with general-purpose digital processors, although special-purpose processors are often used.
- The target data from individual observations is then combined to generate target tracks and other outputs. This phase is often called data processing. The data rates here are in the range 10^6 to 10^9 OPS, which are easily met by general-purpose processors.

Software for digital signal processing requires efficient algorithms for performing the processing functions. The more demanding algorithms include:

- Finite impulse response (FIR) filters.
- Auto- and cross-correlation.
- Fast Fourier transforms (FFT).
- Matrix inversions and other matrix operations.

3 | Radar Performance

3.1 Radar Cross Section (RCS)

The radar cross section (RCS) of a target is the ratio of the power (per unit solid angle) scattered by the target in the direction of the radar receive antenna to the power density (per square meter) incident on the target. RCS, usually represented by σ, has the dimension of area, is usually specified in square meters, and is often described in decibels relative to a one square meter, indicated by dBsm or dBm^2 (Sec. 6.3).

Target RCS depends on the target configuration (shape and materials), radar frequency and polarization, and radar viewing angle. While RCS is generally related to target size, targets having flat surfaces can produce large specular returns, while stealth techniques such as shaping, radar absorbing material (RAM), and non-metallic materials can significantly reduce RCS (Table 3.1).

Target RCS can be obtained from measurements or from detailed computer models. Both approaches are costly, require detailed target descriptions, and generate large amounts of data for frequency and viewing-angle variations.

RCS models. Useful characterization of target RCS is obtained from simpler models. These can be described in three regimes, depending on

TABLE 3.1 Representative Radar Cross Sections for Common Target Types

Target Type	RCS, m^2	RCS, dBsm
Insect or bird	10^{-5} to 10^{-2}	-50 to -20
Man	0.5 to 2	-3 to 3
Small aircraft	1 to 10	0 to 10
Large aircraft	10 to 100	10 to 20
Car or truck	100 to 300	20 to 25
Ship	200 to 1,000	23 to 30

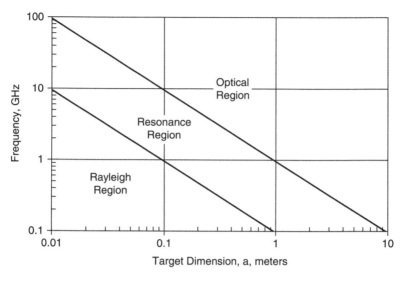

Fig. 3.1   Radar Cross Section Regimes

target size relative to radar wavelength λ, (Fig. 3.1), [17]:

- Rayleigh region. In the region where the target dimension, a, is much less than the signal wavelength, the target RCS has little aspect-angle dependence and varies with the fourth power of frequency. The upper boundary for a in this region is often taken as λ/π:

$$\sigma \propto f^4 \qquad \sigma \propto \lambda^{-4} \qquad (a < \lambda/\pi) \qquad (3.1)$$

- Resonance region. In this region, the wavelength is comparable to the target dimension. The RCS fluctuates with frequency with variations as large as 10 dB, and discontinuities in target shape can produce RCS variations with aspect angle.
- Optical region. In this region, the target dimension is large compared with the signal wavelength. The lower boundary for a is often taken as an order of magnitude greater than the limit for the Rayleigh region:

$$a > 10\lambda/\pi \qquad (3.2)$$

In this region, the RCS of simple shapes can approach their optical cross section. However, more complex targets become a collection of scatterers, which combine coherently to form the target RCS.

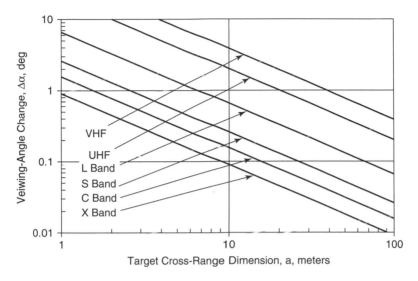

Fig. 3.2 Minimum viewing-angle change required for independent observations vs. target cross-range dimension

For targets in the optical region, aspect angle changes due to target or radar motion cause changes in the distances between scatterers, resulting in a change in the RCS. The change in aspect angle, $\Delta\alpha$, required to produce an independent or uncorrelated RCS observation is:

$$\Delta\alpha \geq \frac{\lambda}{2a} \tag{3.3}$$

In this case, a is the target cross-range dimension (Fig. 3.2).

Similarly, when the signal frequency changes, the distances in wavelength between scatterers also changes. The change in frequency, Δf, required to produce an independent or uncorrelated RCS observation is:

$$\Delta f \geq \frac{c}{2a} \tag{3.4}$$

In this case, a is the target range dimension (Fig. 3.3), [5, Ch. 3].

Swerling RCS fluctuation models. The Swerling target RCS fluctuation models use two bounding cases for RCS decorrelation, [18]:

- Correlated during a dwell. This case assumes that all pulse returns during a dwell or observation of the target are correlated, and that the pulse returns during the next observation period, while also

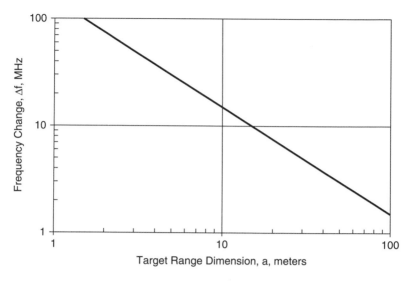

Fig. 3.3   Minimum frequency change required for independent observations vs. target range dimension

correlated, will have a value that is independent from the first group. This might be the case for a rotating search radar (Sec. 3.4), where the returns as the beam sweeps by the target are correlated, but on the successive scans, have different values. For this reason, this case is often called scan-to-scan decorrelation.

- Pulse-to-pulse decorrelation. This case assumes that all pulse returns during a dwell, and also during successive, dwells are decorrelated.

The fluctuations in RCS produced by changes in viewing aspect and signal frequency can be characterized by a probability density function (PDF). A number of PDFs have been used for various target types [18]. The two most common are those proposed by Swerling:

-

$$p(\sigma) = \frac{1}{\sigma_{av}} \exp\left(-\frac{\sigma}{\sigma_{av}}\right) \tag{3.5}$$

where σ_{av} is the average RCS value. This PDF models targets that have many scatterers of comparable magnitude. The exponential PDF results from signal voltage fluctuations that have a Rayleigh PDF, and these targets are called Rayleigh targets.

TABLE 3.2 Swerling Target RCS Fluctuation Models

Probability Density Function and Target Type	Correlated During a Dwell	Pulse-to-Pulse Decorrelation
$p(\sigma) = \dfrac{1}{\sigma_{av}} \exp\left(-\dfrac{\sigma}{\sigma_{av}}\right)$ Many comparable scatterers	Swerling 1	Swerling 2
$p(\sigma) = \dfrac{4\sigma}{\sigma_{av}^2} \exp\left(-\dfrac{2\sigma}{\sigma_{av}}\right)$ One dominant and many small scatterers	Swerling 3	Swerling 4
Non-fluctuating	Swerling 0 or Swerling 5	

-

$$p(\sigma) = \frac{4\sigma}{\sigma_{av}^2} \exp\left(-\frac{2\sigma}{\sigma_{av}}\right) \tag{3.6}$$

This PDF models targets having one dominant scatterer and many small scatterers. It represents a Rayleigh target observed with dual diversity.

The four Swerling target RCS fluctuation models use combinations of these two decorrelation modes and the two PDFs. Non-fluctuating targets are sometimes said to be Swerling 0 or Swerling 5 targets (Table 3.2).

Specular scattering. Scattering from flat or curved metallic surfaces on targets can produce large RCS values that for monostatic radars are in directions normal to the specular surface. The specular return from a flat plate with area A_P is [19]:

$$\sigma = 4\pi \frac{A_P^2}{\lambda^2} \tag{3.7}$$

The angular width θ_S of the specular lobe is about:

$$\theta_S = \frac{\lambda}{W_S} \tag{3.8}$$

where W_S is the dimension of the plate in the plane of the angle.

The specular return normal to a cylinder with radius R_S and length W_S is [20]:

$$\sigma = 2\pi \frac{R_S W_S^2}{\lambda} \tag{3.9}$$

The width of the angular lobe in the plane of the cylinder length is given by Eq. 3.8.

Corner reflectors, also called retro reflectors, consist of three plane surfaces oriented at right angles to each other. They are often used for radar testing and calibration. For monostatic radars, the incident radiation is redirected to the radar, producing a large RCS:

$$\sigma \approx 4\pi \frac{A_{CR}}{\lambda^2} \qquad (3.10)$$

where A_{CR} is the projected area of the corner reflector [21].

Polarization. The scattering properties of a target depend on both the transmitted and received polarizations. For linear polarization, the signal scattered from a target usually has predominantly the same polarization as that incident on the target, and radars usually transmit and receive the same linear polarization. Targets having significant linear structures will usually have larger RCS for polarization aligned with those structures. When a radar receives the orthogonal linear polarization to that transmitted, sometimes called cross polarization, the RCS is usually less, but its value can convey information about the target configuration.

For circular polarization, the received signal is usually predominantly the opposite-sense circular polarization, and radars using circular polarization usually receive opposite-sense polarization to that transmitted (Sec. 2.1). Receiving same-sense circular polarization may convey information on target characteristics. Spherical targets return only opposite-sense polarization. Radars can reject spherical targets, such as rain drops, by receiving same sense polarization at the cost of reduced return from non-spherical targets (Sec. 4.5), [17].

Stealth. Techniques used to reduce RCS include:

- Shaping. Metallic targets reradiate all intercepted energy in some direction. Shaping can reduce the scattering in the direction of the radar, while increasing energy scattered in other directions.
- Radar absorbing material (RAM). Metallic surfaces can be coated with materials that absorb RF energy, reducing reflected energy.
- Non-metallic materials. Materials having dielectric constants much less than that of metals allow signals to partially penetrate target elements, reducing reflections.

Bistatic RCS. Bistatic RCS for a target is comparable to monostatic RCS for that target, except for the forward scatter geometry discussed below. The bistatic angle, β, is the angle at the target formed by the transmitter viewing angle and the receiver viewing angle ($\alpha_T + \alpha_R$ in Fig. 1.2). For reasonably smooth targets and moderate bistatic angles, the bistatic RCS is equal to the monostatic RCS in the direction of the bisector of the bistatic angle, measured at a frequency reduced by $\cos(\beta/2)$.

For bistatic angles near 180 degrees, the forward scatter RCS can become very large, and is given by Eq. 3.7, where A_P here is the projected target area. As was the case with specular RCS, this enhanced forward scatter enhancement occurs for the narrow angular segment given by Eq. 3.8.

3.2 Signal-to-Noise Ratio

The radar range equation. The signal-to noise ratio, (abbreviated S/N or SNR), is a key measure of radar performance, because it determines the radar capability for detecting, measuring and tracking targets. (In addition to noise, interference sources that can affect these radar capabilities include clutter, jamming and chaff. These are discussed in Sec. 4.5 and 4.7.)

The signal-to-noise ratio is defined as the ratio of signal power to noise power at the receiver output. It is calculated by the radar range equation, which relates S/N to key radar and target parameters:

$$\frac{S}{N} = \frac{P_P \, G_T \, \sigma \, A_R \text{PC}}{(4\pi)^2 \, R^4 B k T_S \, L} \tag{3.11}$$

where:

P_P is the peak transmitted power, measured at the transmitter output,

G_T is the transmit antenna gain,

σ is the target radar cross section (RCS),

A_R is the receive antenna effective aperture area,

PC is the signal gain produced when a pulse-compression waveform is used. PC = 1 for constant-frequency pulses (Sec. 5.1),

R is the target range,

B is the radar signal bandwidth,

k is Boltzmann's constant (1.38×10^{-23} J/K), (Sec. 6.2),

T_S is the system noise temperature, measured at the receive antenna output (see below),

L is radar system loss (see below).

Radars normally employ filters matched to the received signal because these matched filters maximize the S/N. A matched filter has a frequency response that is the complex conjugate of the received signal, and therefore has the same signal bandwidth as that signal. When a matched filter is used, the S/N is equal to the ratio of the signal energy at receiver input to the noise power per unit of bandwidth:

$$\frac{S}{N} = \frac{P_P \, \tau \, G_T \, \sigma \, A_R}{(4\pi)^2 \, R^4 \, k \, T_S \, L} \tag{3.12}$$

where τ is the waveform duration. This form of the range equation shows the direct dependence of S/N on the waveform energy, $P_P\tau$.

When the receive gain, G_R is used instead of the receive effective aperture area, A_R, these equations become:

$$\frac{S}{N} = \frac{P_P \, G_T \, \sigma \, G_R \, \lambda^2 \, P \, C}{(4\pi)^3 \, R^4 \, B \, k \, T_S \, L} \tag{3.13}$$

$$\frac{S}{N} = \frac{P_P \, \tau \, G_T \, \sigma \, G_R \, \lambda^2}{(4\pi)^3 \, R^4 \, k \, T_S \, L} \tag{3.14}$$

When $G_T = G_R$, then G^2 can be used in Eq. 3.13 and 3.14.

In the above equations, R is the monostatic radar range. For bistatic radars, substitute $R_T^2 R_R^2$ for R^4, where R_T is the range from the transmit antenna to the target and R_R is the range from the target to the receive antenna, (Fig. 1.2).

For CW radars, B is the receiver integration bandwidth, and $\tau = 1/B$ is the receiver integration time.

System Loss. Losses that should be included in the system loss factor, L, include:

- Transmit microwave loss from the transmitter output (or point where transmit power is measured), to the transmit antenna
- Two-way propagation losses from the transmit antenna to the target and the target to the receive antenna. These can include losses from the atmosphere (Sec. 4.1), rain (Sec. 4.2), and the ionosphere (Sec. 4.6).
- Two-way off-broadside scan loss for phased array antennas (Sec. 2.1).
- Signal-processing losses, including departure from an ideal matched filter, quantization error, integration loss, straddling range or Doppler bins, signal distortion, and CFAR setting loss.

Transmit and receive antenna loses, including ohmic loss and aperture efficiency loss, are normally included in the corresponding gain and effective aperture area parameters. If they are not, they should be included in the system loss. Any microwave loss from the receive antenna to the point where receiver noise temperature is measured should also be included in the system loss.

When the individual losses are specified as power ratios, they should be multiplied to give the system loss as a power ratio. When they are specified in dB, they should be added to give the system loss in dB.

System Noise Temperature. The system noise temperature is conventionally defined at the output of the receive antenna. It is the sum of four components, each adjusted to the receive antenna output:

- The noise seen by the antenna. For a ground-based radar, the portions of the antenna pattern viewing the earth's surface are usually assumed to see a temperature of 290 K. Portions of the pattern viewing the sky see a combination of noise from atmospheric absorption and cosmic radiation. The sky temperature is in the range 10 K to 100 K for radar frequencies in the 1–10 GHz range [22, Ch. 8]. Considering that even antennas viewing the sky will have sidelobes viewing the earth and vice versa, the resulting noise temperature, T_A, will lie between these values, typically 200 K. This temperature is reduced by the receive antenna ohmic and aperture efficiency losses, and its contribution to T_S is $T_A/(L_O + L_E)$.

- The noise from receive antenna ohmic loss is equal to the amount this loss exceeds unity times the temperature at which it occurs, usually assumed to be 290 K. This is also reduced by the antenna ohmic and aperture efficiency losses, and its contribution to T_S is $290(L_O - 1)/L_O L_E$. Any microwave loss from the antenna to the point where the system noise temperature is defined should be included with the antenna ohmic loss.

- Noise from the microwave line from the receive antenna (or the point where the system noise temperature is specified), to the receiver, L_M, is equal to the amount this loss exceeds unity times the temperature at which it occurs, usually assumed to be 290 K. Its contribution to T_S is $290(L_M - 1)$.

- The receiver noise, (Sec. 2.3), is increased by the microwave loss factor. Its contribution to T_S is $T_R L_M$.

The resulting system noise temperature, defined at the receive antenna output, is:

$$T_S = \frac{T_A}{L_O \, L_E} + \frac{290 \, (L_O - 1)}{L_O \, L_E} + 290 \, (L_M - 1) + T_R \, L_M \qquad (3.15)$$

In radars that have low antenna and microwave losses, Eq. 3.15 reduces to:

$$T_S \approx T_A + T_R \qquad \text{(for } L_O, \, L_E, \text{ and } L_M \approx 1) \qquad (3.16)$$

This will often be the case for phased arrays using T/R modules.

Pulse integration. The S/N can be increased by combining signal returns from two or more transmitted pulses using pulse integration. Pulse integration may be coherent or noncoherent (Fig. 3.4).

With coherent integration, also called predetection integration, the pulse returns are added as RF, IF or coherently detected signals. This requires:

- Coherent radar operation.
- A stable target that produces coherent signal returns. This requirement is met by Swerling 1 and 3 targets, as well as by nonfluctuating targets.

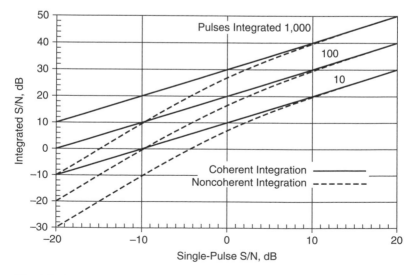

Fig. 3.4 Integrated S/N for coherent and noncoherent integration vs. single-pulse S/N (made using custom radar functions in [5])

- A stable propagation path. This is usually met, except at low frequencies where the ionosphere can produce signal fluctuations (Sec. 4.6).

Coherent integration increases S/N by the number of pulses coherently integrated:

$$\left(\frac{S}{N}\right)_{CI} = n\,\frac{S}{N} \tag{3.17}$$

where $(S/N)_{CI}$ is the coherently integrated signal-to noise ratio, n is the number of pulsed coherently integrated, and S/N is the single-pulse signal-to-noise ratio.

The Doppler-frequency shift for targets with non-zero radial velocity will produce a linear pulse-to-pulse phase shift that prevents successive pulses from adding in phase. If the radial velocity is known, the signal processing can compensate for these phase shifts. When the radial velocity is not known, the processor can consider the range of possible radial velocities, or pulse-Doppler processing may be employed (Sec. 5.3).

With noncoherent integration, also called post-detection integration and sometimes incoherent integration, pulse returns are added after detection or demodulation in the radar receiver, where they no longer retain phase information. These signals have experienced a detection loss, L_D, [10, Ch. 3]:

$$L_D = \frac{1 + \frac{S}{N}}{\frac{S}{N}} \tag{3.18}$$

When n pulses are noncoherently integrated, the resulting signal-to-noise ratio, $(S/N)_{NI}$, is:

$$\left(\frac{S}{N}\right)_{NI} = \frac{n\,\left(\frac{S}{N}\right)^2}{1 + \left(\frac{S}{N}\right)} \tag{3.19}$$

Since noncoherent integration takes place after detection, it can be performed on any target; fluctuating or non-fluctuating. There is no requirement for frequency or phase stability.

Range changes due to target radial velocity can result in the pulses not fully adding when the range change over the integration period is greater than the waveform range resolution (Sec. 5.1). This situation is referred to as range walk. It can be compensated for during signal processing and requires much less accuracy in radial velocity than compensating for phase change in coherent integration.

Since the detection loss is significant at low S/N, noncoherently integrating many low-S/N pulses to achieve a high integrated signal-to-noise ratio is inefficient. Noncoherent integration gain is always lower than coherent integration gain. However, detection of fluctuating targets can be enhanced by noncoherently integrating a small number of pulses (Sec. 3.3).

3.3 Detection

Radar detection is the process of examining received radar signals to determine if a target is present in background noise. Both the noise background and the target signals have random properties, so detection is a statistical process.

The detection approach usually used is to set a threshold that excludes most noise samples. Signals that exceed the threshold are declared to be detections, while target signals that fail to exceed the threshold are not detected. Noise samples that exceed the threshold are called false alarms, The detection performance is then characterized by probability of detection P_D and probability of false alarm P_{FA}. The detection performance depends on the characteristics of the signal and noise background, and on S/N.

False Alarms. Background noise is usually assumed to be white Gaussian noise, which has a flat spectrum and a Rayleigh PDF for its voltage, and exponential PDF for its power.

For detection with a single observation, P_{FA} is given by:

$$P_{FA} = e^{-Y_T} \qquad (3.20)$$

where Y_T is the threshold power level, normalized to the mean noise power. The threshold level can be set to provide the desired P_{FA}:

$$Y_T = \ln\left(\frac{1}{P_{FA}}\right) \qquad (3.21)$$

When n observations are used for detection, there is no closed-form expression for P_{FA}, and the desired threshold level is found by recursive calculation.

The receiver produces independent noise samples at the rate of the noise bandwidth. With matched-filter processing, this is normally equal to the signal bandwidth B, and the reciprocal of the compressed pulse duration (Sec. 5.1). For continuous receiver operation, the false-alarm rate, r_{FA}, is:

$$r_{FA} = P_{FA}B \qquad (3.22)$$

and the average time between false alarms, t_{FA}, is:

$$t_{FA} = \frac{1}{r_{FA}} = \frac{1}{P_{FA}B} \qquad (3.23)$$

The threshold level is set to produce an acceptable false-alarm rate, either based on calculation or on feedback from measurement of the noise level. Constant-false-alarm rate (CFAR) techniques automatically adjust the threshold as the noise level varies, based on feedback from the false-alarm rate that is being experienced. These adjustments are critical when the noise level is raised dramatically by jamming (Sec. 4.7).

Single-pulse detection. Calculation of P_D involves integration of a complex function of the signal-plus-noise and signal-to-noise values of the observations. Computational techniques and approximations are available in the literature [23, 6, Ch. 2). The Swerling RCS models, (Sec. 3.1), are most widely used for detection calculations. These result in a distribution of S/N about a mean value that corresponds to the average RCS value. Graphical results are presented here for the Swerling target-signal fluctuation models, and for non-fluctuating targets.

For a single observation, the statistics for Swerling 1 and Swerling 2 targets are the same, and the statistics for Swerling 3 and Swerling 4 targets are the same. The required average single-pulse S/N is plotted as a function of the desired P_D for Swerling 1 and 2, Swerling 3 and 4, and non-fluctuating targets in Figs. 3.5, 3.6, and 3.7 for P_{FA} values of 10^{-4}, 10^{-6}, and 10^{-8} respectively.

Detection with coherent integration. Coherent integration requires a coherent radar and a stable target return. The latter is provided by Swerling 1 and Swerling 3 targets, as well as by non-fluctuating targets (Sec. 3.1). Coherent integration can not be used with targets having pulse-to-pulse fluctuations such as Swerling 2 and Swerling 4 targets. Noncoherent integration (see below), can be used with these.

Coherently integrating a group of pulse returns produces a single target observation. The detection statistics are the same as for a single pulse, only using the coherently-integrated S/N, taking into account any associated signal-processing loss. The coherently-integrated S/N needed to provide a desired P_D can be found from Figs. 3.5, 3.6, and 3.7 for Swerling 1, Swerling 3, and non-fluctuating targets.

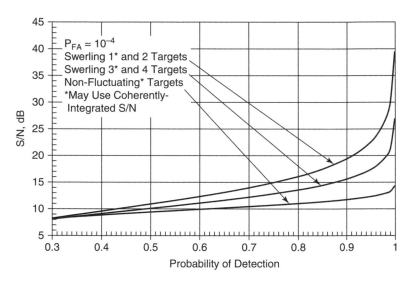

Fig. 3.5   Single-pulse and coherently integrated S/N vs. P_D for $P_{FA} = 10^{-4}$ (made using the custom radar functions in [5])

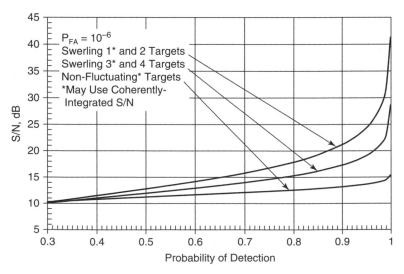

Fig. 3.6   Single-pulse and coherently integrated S/N vs. P_D for $P_{FA} = 10^{-6}$ (made using the custom radar functions in [5])

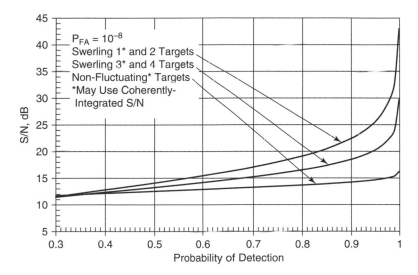

Fig. 3.7   Single-pulse and coherently integrated S/N vs. P_D for $P_{FA} = 10^{-8}$ (made using the custom radar functions in [5])

Detection with noncoherent integration. Noncoherent integration can be used with both fluctuating and non-fluctuating targets:

- With stable (non-fluctuating) targets, noncoherent integration is less efficient than coherent integration due to the detection loss (Sec. 3.2), and as the number of pulses integrated, n, increases, the required single-pulse S/N decreases more slowly than 1/n. This is the case for Swerling 1, Swerling 3, and non-fluctuating targets.
- With fluctuating targets, the required single-pulse S/N initially decreases more rapidly than 1/n, producing a range of n values where the single-pulse S/N with noncoherent integration is less than that for coherent integration. This is a consequence of the uncorrelated signal returns, which are unlikely to all have low values during a dwell. The minimum pulse S/N is obtained by noncoherently integrating a number of pulses, n, in the range 5 to 20. The effect is more pronounced for higher values of P_D, and for the more highly-fluctuating Swerling 2 targets than for Swerling 4 targets.

Figs. 3.8 through 3.22 show the single-pulse S/N as a function of the number of pulses coherently integrated for P_D values of 0.999, 0.99. 0.95, 0.9, and 0.7, and values of P_{FA} of 10^{-4}, 10^{-6}, and 10^{-8}.

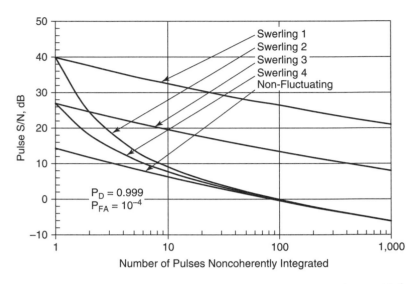

Fig. 3.8   Pulse S/N vs. number of pulses integrated for $P_D = 0.999$ and $P_{FA} = 10^{-4}$ (made using the custom radar functions in [5])

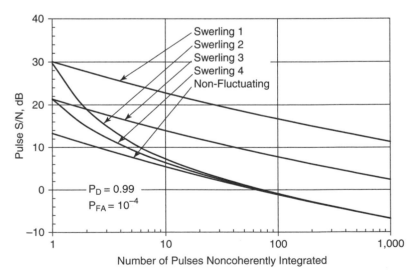

Fig. 3.9   Pulse S/N vs. number of pulses integrated for $P_D = 0.99$ and $P_{FA} = 10^{-4}$ (made using the custom radar functions in [5])

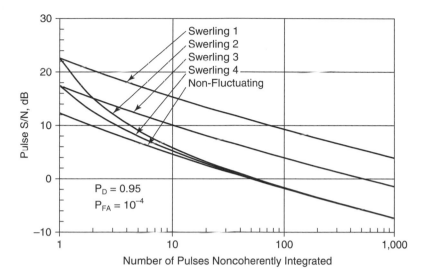

Fig. 3.10 Pulse S/N vs. number of pulses integrated for $P_D = 0.95$ and $P_{FA} = 10^{-4}$ (made using the custom radar functions in [5])

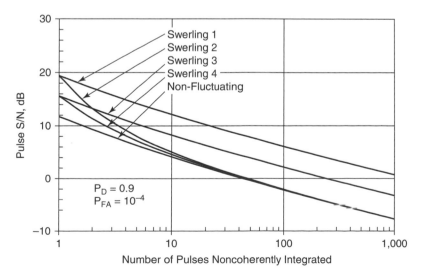

Fig. 3.11 Pulse S/N vs. number of pulses integrated for $P_D = 0.9$ and $P_{FA} = 10^{-4}$ (made using the custom radar functions in [5])

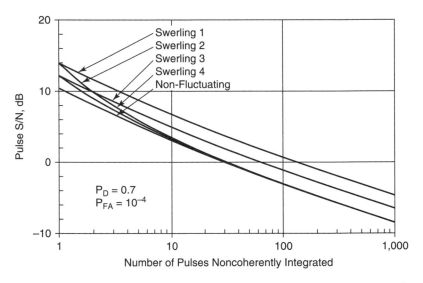

Fig. 3.12 Pulse S/N vs. number of pulses integrated for $P_D = 0.7$ and $P_{FA} = 10^{-4}$ (made using the custom radar functions in [5])

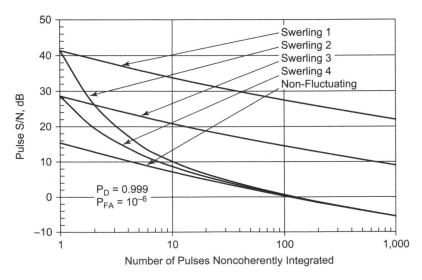

Fig. 3.13 Pulse S/N vs. number of pulses integrated for $P_D = 0.999$ and $P_{FA} = 10^{-6}$ (made using the custom radar functions in [5])

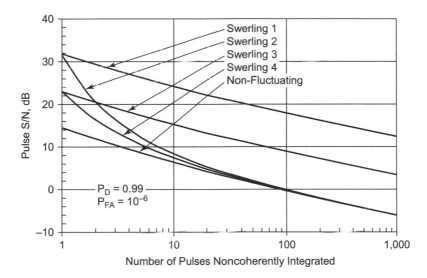

Fig. 3.14   Pulse S/N vs. number of pulses integrated for $P_D = 0.99$ and $P_{FA} = 10^{-6}$ (made using the custom radar functions in [5])

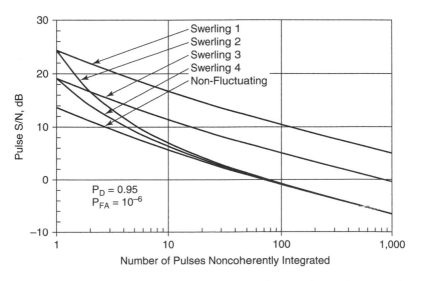

Fig. 3.15   Pulse S/N vs. number of pulses integrated for $P_D = 0.95$ and $P_{FA} = 10^{-6}$ (made using the custom radar functions in [5])

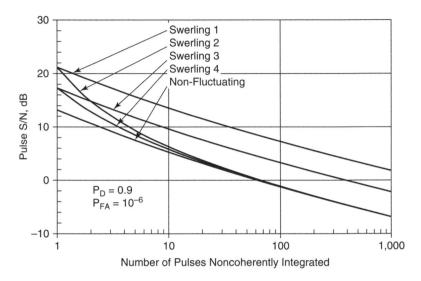

Fig. 3.16   Pulse S/N vs. number of pulses integrated for $P_D = 0.9$ and $P_{FA} = 10^{-6}$ (made using the custom radar functions in [5])

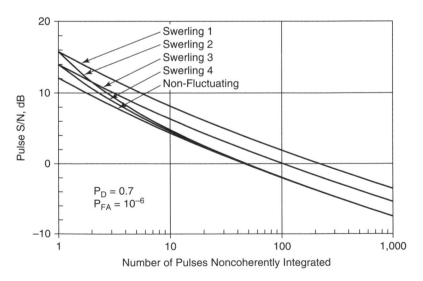

Fig. 3.17   Pulse S/N vs. number of pulses integrated for $P_D = 0.7$ and $P_{FA} = 10^{-6}$ (made using the custom radar functions in [5])

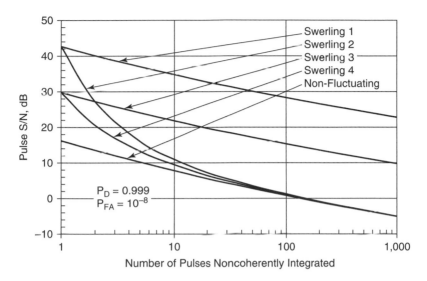

Fig. 3.18   Pulse S/N vs. number of pulses integrated for $P_D = 0.999$ and $P_{FA} = 10^{-8}$ (made using the custom radar functions in [5])

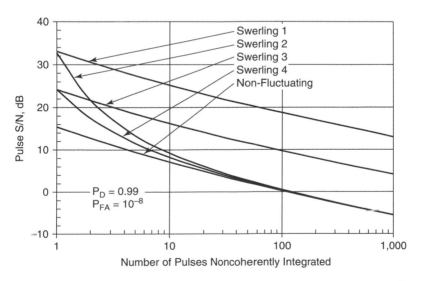

Fig. 3.19   Pulse S/N vs. number of pulses integrated for $P_D = 0.99$ and $P_{FA} = 10^{-8}$ (made using the custom radar functions in [5])

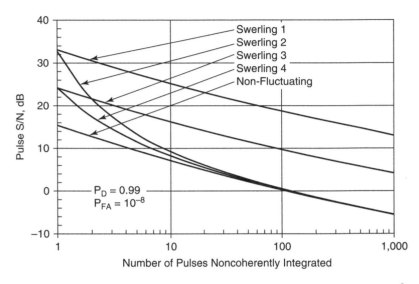

Fig. 3.20 Pulse S/N vs. number of pulses integrated for $P_D = 0.95$ and $P_{FA} = 10^{-8}$ (made using the custom radar functions in [5])

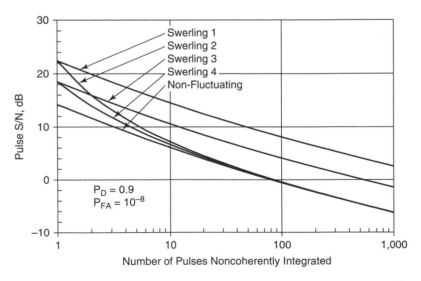

Fig. 3.21 Pulse S/N vs. number of pulses integrated for $P_D = 0.9$ and $P_{FA} = 10^{-8}$ (made using the custom radar functions in [5])

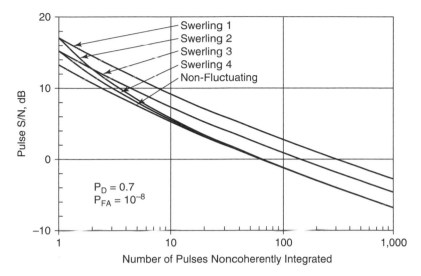

Fig. 3.22 Pulse S/N vs. number of pulses integrated for $P_D = 0.7$ and $P_{FA} = 10^{-8}$ (made using the custom radar functions in [5])

The advantages of noncoherent integration in reducing the pulse S/N needed for high-P_D detection can be obtained with stable targets by using frequency changes that meet the requirement of Eq. 3.4. This essentially converts Swerling 1 targets to Swerling 2, and Swerling 3 targets to Swerling 4. When large numbers of pulses are used for detection of stable targets, it may be advantageous to integrate groups of pulse coherently, producing 5 to 20 groups of coherently-integrated pulses. Then use frequency changes among these groups in order to operate in the range of noncoherent observations that provides the lowest S/N requirement.

The Swerling models address only fully correlated dwells (Swerling 1 and 3), and fully uncorrelated target observations (Swerling 2 and 4). Cases where targets are partially correlated lie between these extremes. Barton [6, Ch 2], gives an empirical method for calculating the required single-pulse S/N for these cases.

Cumulative detection. Another technique for detection using multiple pulses is cumulative detection. A series of n pulses or integrated dwells is transmitted on the target. Detection is declared if the return from one pulse or dwell exceeds the threshold. Cumulative detection is a special case of m-out-of-n detection, sometimes called binary integration, which sets the threshold to require m detections out of n pulses or integrated dwells.

For targets with pulse-to-pulse fluctuations, such as Swerling 2 and 4, the cumulative detection probability is:

$$P_D = 1 - (1 - P_{DO})^n \qquad (3.24)$$

where P_{DO} is the detection probability for a single observation, (pulse or dwell). The value of P_{DO} needed to provide a given P_D is:

$$P_{DO} = 1 - (1 - P_D)^{1/n} \qquad (3.25)$$

To maintain the false-alarm probability P_{FA} for the series of n observations, the false-alarm probability for each observation, P_{FAO}, must be reduced to:

$$P_{FAO} = \frac{P_{FA}}{n} \qquad (3.26)$$

The required single-observation S/N with cumulative detection initially decreases more rapidly than $1/n$, producing a range of n values where it is less than that for coherent integration. Cumulative detection requires higher observation, (pulse or dwell), S/N than noncoherent integration. However, it is easy to implement, the observations can be widely spaced in time, and range walk need not be considered [5, Ch, 6].

3.4 Search

Radar search (also called surveillance) examines a volume of space to detect targets in that volume. The search objectives are:

- Search volume, usually specified as a solid angle, ψ_S, a maximum range, and sometimes a minimum range (see Eq. 1.6).
- Target characteristics, including target RCS, RCS fluctuation characteristics, and velocity.
- Detection parameters, including P_D and P_{FA}.

Search Equation. The equation that describes radar performance in search is:

$$R_D = \left[\frac{P_A\, A_R\, t_S\, \sigma}{4\pi\, \psi_S\, n\, (S/N)\, k\, T_S\, L_S} \right]^{1/4} \qquad (3.27)$$

where:

t_S is the search time.

ψ_S is the solid angle searched.

n is the number of pulses integrated in the search observation.

S/N is the pulse S/N that, along with n, provides the required P_D and P_{FA}(Sec. 3.2).

L_S is the search loss.

The other parameter are defined in Sec. 3.2.

The search loss factor, L_S, includes [5, Ch. 7]:

- The radar system loss, L, (Sec. 3.2). Here the off-broadside scan loss and the propagation loss are average values over the solid angle searched.
- Beamshape loss, which accounts for targets that are not at the beam peak. This factor is typically in the range of 1.6 to 2.5 dB, (factor 1.45 to 1.78).
- Beam-packing loss that results from overlap of circular (or elliptical) beams in the search pattern. A value of 0.8 dB (factor 1.2) is typical.
- Signal processing losses including integration loss when pulse integration is used, and any other processing losses from the search mode.
- Losses due to non-ideal distribution of radar energy in the search mode.

When only the radar parameters in Eq. 3.27 are considered:

$$R_D \propto \left[\frac{P_A \, A_R}{T_S \, L_S} \right]^{1/4} \tag{3.28}$$

This allows comparison of the search performance of radars. In most radars, T_S and L_S vary over a small range. Then, the radar search performance is approximately proportional to the fourth root of $P_A \, A_R$, which is called the power-aperture product (Fig. 3.23). In multi-function radars, P_A here is the portion of the radar average power used in the search mode.

The radar search performance is independent of the transmit antenna gain. Increasing the transmit gain would increase the radar sensitivity, but it also narrows the beamwidth. This requires more beam positions to be searched, allowing less energy in each beam, canceling out the increase in sensitivity.

Equation 3.27 shows that increasing the search time increases the detection range. However the search time may be constrained by target dynamics, among other factors. When a target is approaching the

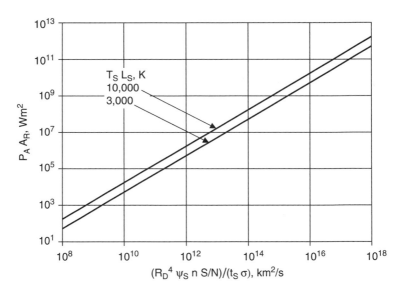

Fig. 3.23 Power-aperture product vs. search parameters

radar with radial velocity V_R, the search time that maximizes the assured detection range, R_A, is:

$$t_S = \frac{R_D}{4 V_R} \quad \text{(for maximum } R_A\text{)} \tag{3.29}$$

When this search time is used:

$$R_D = \left[\frac{P_A A_R \sigma}{16\pi V_R \psi_S n (S/N) k T_S L_S} \right]^{1/3} \quad \text{(for maximum } R_A\text{)} \tag{3.30}$$

Then the assured detection range is then:

$$R_A = \frac{3}{4} R_D \quad \text{(maximum)} \tag{3.31}$$

Rotating search radars. These radars employ parabolic-reflector or array antennas that have narrow azimuth beamwidth. The antenna rotates continuously in azimuth, usually at a constant rate, providing 360-degree search coverage with a rotation period typically 10 to 20 seconds. The elevation-angle coverage is tailored to match the targets of interest. Maximum gain is usually provided at low elevation angles. The gain is often reduced at higher elevation angles to provide coverage only to maximum aircraft altitudes, typically 10 to 15 km.

The number of pulses that observe the target as the beam scans past is:

$$n = \frac{\theta_A \, t_R \, \text{PRF}}{2\pi} \tag{3.32}$$

where θ_A is the azimuth beamwidth, t_R is the rotation period, and PRF is the pulse repetition frequency, which is assumed to be fixed. These pulses are usually integrated non-coherently or coherently depending on the target fluctuations and the radar capability. Target returns are usually uncorrelated from scan to scan, so when the pulses within a dwell are correlated, Swerling 1 or 3 may be used, and when the pulses in a dwell are uncorrelated, Swerling 2 or 4 may be used.

The detection range for rotating search radars can be found using Eq. 3.27, with search time, $t_S = t_R$, and search solid angle ψ_S equal to 2π times the elevation-angle coverage. Normally only the high-gain used for low elevation-angle coverage should be used here, since the power needed for the reduced-range high-angle coverage is typically only about 50 percent that used at the low elevation angles. This can be taken into account by a loss of about 2 dB. The beam-shape loss is about 1.6 dB. If n is less than 6, there is a small scanning loss to account for beam motion between transmission and reception [5, Ch. 7].

Alternatively, Eq. 3.11, 3.12, 3.13, or 3.14 can be used to find the detection range, with S/N in these equations replaced with n times the pulse S/N required for detection, and the search losses included.

Some rotating search radars have stacked elevation beams to provide the high elevation-angle coverage and to allow measurement of elevation angle (Sec. 3.5). One approach uses a fan beam for transmit, and stacked elevation beams for receive. Another approach transmits and receives at several elevation angles as the antenna rotates in azimuth.

Phased-array volume search. Phased-array radars normally have narrow, pencil beams that can be rapidly repositioned by electronic steering. They perform search by transmitting and receiving in a sequence of beam positions, usually arranged in a grid covering the search solid angle. Full-field-of-view (FFOV), phased arrays can scan to angles about 60 degrees off broadside. It takes three or more of such arrays to provide hemispheric coverage. Limited-field-of-view (LFOV), phased arrays provide less angular coverage and are used to search smaller angular sectors (Sec. 2.1).

The beam positions in a search grid are typically spaced by about a beamwidth. This produces a beam-packing loss of about 0.8 dB, and a beamshape loss of about 2.5 dB. The scan loss varies with off-broadside scan angle, requiring variable energy in individual beam positions if S/N is to be maintained. An average scan-loss factor is often used in evaluations [5, Ch.7].

Since the phased-array beamwidth increases with $\cos^{-1}$ of the scan angle off broadside, the beam spacing and number of beams used to cover the search solid angle must be adjusted accordingly. Laying out the search pattern in sine space, where the beam pattern in invariant with scan angle, simplifies this task [24].

Equation 3.27 can be used to find the search detection range, incorporating the losses mentioned above. The number of pulses per beam position, n, is:

$$n = \frac{t_S \, \text{PRF}}{n_B} \tag{3.33}$$

where n_B is the number of beams in the search pattern, and the PRF is assumed constant.

Coherent or noncoherent integration can be used for the dwell in each beam position, depending on the target-signal correlation. For long search ranges where PRF is low, and for large values of n_B, the number of pulses per beam position may fall below an efficient number for noncoherent integration (Sec. 3.3). In such cases, pulses may be transmitted in several beam positions in rapid succession, and multiple receive beams used to receive signal returns.

Barrier search. In barrier search, a radar searches a angular segment narrow in one dimension, (elevation angle for this discussion), and wide in the other dimension, (azimuth for this discussion). The objective is to detect targets as they pass through the narrow barrier. Common applications are:

- Horizon search, where the radar searches an azimuth sector at low elevation angle to detect artillery shells or missiles as they come above the horizon.
- Push-broom search, where an airborne radar searches side-to side to detect targets as it passes over them.

To assure that targets are detected as they pass through the barrier, the minimum barrier search time is:

$$t_S = \frac{R_T \, \phi_E}{V_T} \tag{3.34}$$

where:

- R_T is the range to the target. The shortest target range of interest is used to give a value of t_S that is small enough to assure detection
- ϕ_E is the elevation-angle dimension of the search pattern. When a single row of beam positions is used for the barrier, ϕ_E is equal to the elevation beamwidth.
- V_T is the target velocity component in the elevation-angle direction, normal to the radar line-of-sight.

If the azimuth search angle is ϕ_A, the search solid angle is:

$$\psi_S = \phi_A \, \phi_E \qquad (3.35)$$

The detection range, R_D, can be found from Eq. 3.27, using the values from Eqs. 3.34 and 3.35. A beam-packing loss of about 0.8 dB and a beamshape loss of about 2.5 dB should be included in the search losses. The detection range R_D must exceed the target range R_T for target detection.

Phased-array radars can perform barrier search by transmitting one or more rows of beams that meet the above requirements. The considerations discussed earlier for the number of pulses per beam position apply here.

Dish radars may be limited in their barrier-search capability by their maximum azimuth scan velocity and acceleration [5, Ch.7]. Within these limitations, the detection range can be found from Eq. 3.11, 3.12, 3.13, or 3.14, with S/N in these equations replaced with n times the pulse S/N required for detection, and the search losses included.

3.5   Measurement

Radars measure target range and two angular coordinates to determine target position. Coherent radars can also directly measure target radial velocity. A target must be resolved from other targets in at least one coordinate for its parameters to be measured:

- Range. Range resolution, ΔR is:

$$\Delta R = \frac{c}{2\,B} = \frac{t_R\,c}{2} \qquad (3.36)$$

where B is the signal bandwidth, and t_R is the compressed pulse duration and matched-filter processing is used.

- Angle. The angle resolution is defined by the beamwidth, θ. The physical cross-range separation, ΔD, is:

$$\Delta D = R\theta \qquad (3.37)$$

where R is the target range.
- Radial velocity. With coherent radars, the radial-velocity resolution, ΔV resulting from Doppler-frequency shift is:

$$\Delta V = \frac{\lambda}{2\tau} = \frac{\lambda f_R}{2} \qquad (3.38)$$

where τ is the waveform duration, and f_R is the Doppler frequency resolution and matched-filter processing is used.

Measurement errors. Radar measurement errors are classified as:

- Signal-to-noise dependent random errors. These vary inversely with the square root of S/N, and usually dominate the radar measurement error. Either single-pulse, or integrated S/N may be used (Sec. 3.2), The error values are usually assumed to have Gaussian probability density, and when independent measurements are averaged, the error is reduced by the square root of the number of such measurements. The radar parameters that affect S/N are $P_P A_R G_T / T_S L$ (Eq. 3.11 and 3.12). These can be used to compare the measurement performance of radars. When T_S and L vary over a small range, radar measurement errors are approximately proportional to $(P_P A_R G_T)^{-1/2}$.
- Fixed random measurement errors. These can result from noise in the later stages of the radar receiver, or from propagation uncertainties that vary randomly from observation to observation (Sec. 4.3 and 4.6). They set a limit on how far random measurement errors can be reduced by increasing S/N. This is sometimes expressed as a pulse-splitting or a beam-splitting limitation at high S/N. The errors are usually assumed to have Gaussian probability density, and when independent measurements are averaged, the error is reduced by the square root of the number of such measurements.
- Bias errors. These errors result from errors or drift in the radar alignment and calibration, and from propagation effects, (Sec. 4.3 and 4.6). They may vary over long time periods, but are assumed to remain fixed over the time that a target is observed. Since bias errors will be the same for targets in the same general location,

they do not affect relative measurements or radar tracking of these targets, and need not be included in the measurement errors for these situations.

- Errors from interference such as radar clutter (Sec. 4.5) or jamming (Sec. 4.7).
- Errors from multipath (Sec. 4.4), target scintillation, and glint (see below).

The overall measurement error is the combination of these components. When S/N-dependent and fixed random errors both have Gaussian distributions, as is usually assumed, they can be combined on a root-sum-square, (rss), basis. The other errors can be added as appropriate. Sometimes they are all combined using rss, for convenience.

For Gaussian error distributions, it is usual to specify the standard deviation of the error, designated by σ. About 68% of the error values will lie between $\pm\sigma$. When it is desired to use an error bound that includes a greater percentage of measurements, a multiple of σ is used, such as $\pm 3\sigma$, which includes 99.7% of the error values, (Fig. 3.24).

Range Measurement. Range is determined by the time interval between signal transmission and reception (Eq. 1.1 and 1.4). It is often

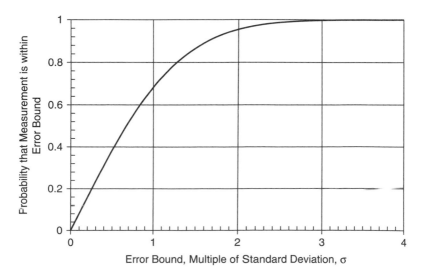

Fig. 3.24 Probability that a measurement is included vs. error bound

measured by estimating the centroid of the detected compressed pulse. Other techniques include comparing the signal levels in early and late gates, and in early radars, observing an oscilloscope display. These techniques have been shown to give comparable results when properly implemented.

The S/N dependent range error, σ_R, usually the dominant range error, is:

$$\sigma_R = \frac{\Delta R}{\sqrt{2\,(S/N)}} = \frac{c}{2\,B\sqrt{2\,(S/N)}} \tag{3.39}$$

where the S/N can be single-pulse, coherently integrated or noncoherently integrated (Sec. 3.2).

Fixed random range errors from internal radar noise are typically 0.05 to 0.0125 of the range resolution, (pulse splitting factor of 20 to 80), (Fig. 3.25). Radar bias errors in range are usually from incorrect compensation for the path length from the antenna to the receiver, and can be made small by careful calibration. Bias errors from atmospheric and ionospheric propagation can be significant. These can be reduced by correcting for their estimated values (Sec. 4.3 and 4.6).

Angle measurement. Angular measurements of signal returns are usually made in two orthogonal planes, often azimuth and elevation angle.

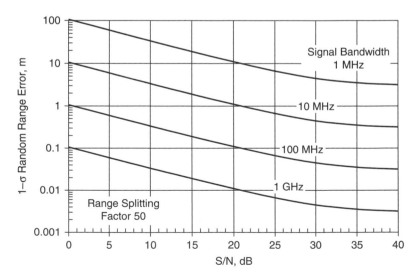

Fig. 3.25   S/N-dependent and fixed random range measurement error vs. S/N

Monopulse antennas are used for angle measurement by many radars (Sec. 2.1). Another technique transmits and receives signals at several angles around the target, as is done in conical scanning. With rotating search radars, target azimuth is estimated from the centroid of the series of pulse returns as the antenna sweeps past the target (Sec. 3.4). These latter methods are subject to errors from target signal fluctuations since, unlike monopulse, they are not performed on the same receive pulse. When fluctuations are small, these methods produce results comparable to those from monopulse measurement.

The S/N dependent angle error, σ_A, usually the dominant angle error, is:

$$\sigma_A = \frac{\theta}{k_M \sqrt{2\,(S/N)}} \approx \frac{\theta}{1.6 \sqrt{2\,(S/N)}} \tag{3.40}$$

Where θ is the beamwidth in the plane of the angle measurement, and k_M is the monopulse antenna pattern difference slope. Typically, $k_M \approx 1.6$. The S/N can be single-pulse, coherently integrated or noncoherently integrated (Sec. 3.2). The cross-range dimensional error, σ_D, is then:

$$\sigma_D = R\,\sigma_A \tag{3.41}$$

In phased-array radars, the beamwidth varies inversely with the cosine of the off-broadside scan angle, and the scan loss increases with off-broadside scan. Both these effects increase σ_A, if they are not compensated for (Eq. 2.12 and 2.13).

Fixed random angle errors from internal radar noise are typically 0.025 to 0.008 of the beamwidth, (beam splitting factor of 40 to 125), (Fig. 3.26). Radar bias errors in angle are usually from errors in knowing the physical alignment of the antenna. For fixed radars, these are usually small. Bias errors from atmospheric and ionospheric propagation can be large, especially for elevation angle at low elevation. These can be reduced by correcting for their estimated values (Sec. 4.3 and 4.6).

In phased arrays, both fixed random and bias errors can have components that are independent of scan angle, and components that vary with the reciprocal of the cosine of the off-broadside scan angle. These may usually be combined by the rss method

In some situations signals from jamming or clutter may affect the beams used for angular measurement differently, creating larger errors than predicted by Eq. 3.40.

Target glint is the effect of scatterers in the target interacting to produce angle errors that can exceed the angular dimensions of the target. This effect is usually confined to short ranges where the target

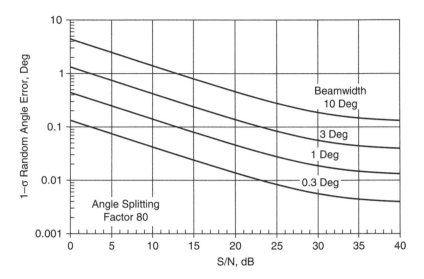

Fig. 3.26   S/N-dependent and fixed random angle measurement error vs. S/N

angular extent is significant; for example with target homing radars or with precision-approach radars.

Radial velocity. Coherent radars may determine target radial velocity by measuring the Doppler-frequency shift of the received signal (Eq. 1.2, 1.3, and 1.5). The resulting S/N dependent radial-velocity error, σ_V, usually the dominant radial-velocity error, is:

$$\sigma_V = \frac{\lambda}{2\,\tau\,\sqrt{2\,(S/N)}} = \frac{\Delta V}{\sqrt{2\,(S/N)}} \tag{3.42}$$

The S/N can be single-pulse, coherently integrated or noncoherently integrated (Sec. 3.2). Fixed random errors from internal noise can limit the radial-velocity measurement accuracy for large values of S/N (Fig. 3.27). Radar bias errors are usually small, and both random and bias errors from propagation are usually small.

Radial velocity can also be measured noncoherently by combining two or more range measurements. The resulting measurement error is:

$$\sigma_V = \frac{\sqrt{2}\,\sigma_R}{t_M} \qquad \text{(2 pulses)} \tag{3.43}$$

$$\sigma_V = \frac{\sqrt{12}\,\sigma_R}{\sqrt{n}\,t_M} \qquad \text{(6 or more pulses)} \tag{3.44}$$

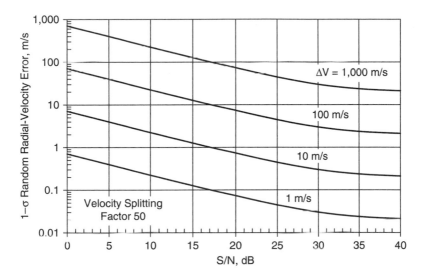

Fig. 3.27 S/N-dependent and fixed random radial-velocity measurement error vs. S/N for coherent measurement

where t_M is the duration of the measurement and n is the number of pulses used. In most cases the radial-velocity measurement error is far less when using coherent Doppler-shift measurement, (Eq. 3.42), than when using noncoherent range measurements (Eq. 3.43 and 3.44), [5, Ch. 8].

Cross-range velocity can be measured using two or more angle measurements. The resulting measurement error is:

$$\sigma_C = \frac{\sqrt{2}\, R\, \sigma_A}{t_M} \qquad \text{(2 pulses)} \qquad (3.45)$$

$$\sigma_C = \frac{\sqrt{12}\, R\, \sigma_A}{\sqrt{n}\, t_M} \qquad \text{(6 or more pulses)} \qquad (3.46)$$

The cross-range velocity error usually far exceeds the radial-velocity error.

3.6   Tracking

Radar tracking combines a series of measurements of target position and radial velocity to estimate the target's path. Tracking provides an estimate of target course and speed, detects target maneuvers, and predicts target position.

Tracking modes. The tracking mode and capabilities of a radar depend on the antenna configuration (Sec. 2.1):

- Rotating surveillance radars measure target position and possibly radial velocity once each rotation period. Tracks are generated by combining these measurements in a mode called track-while-scan (TWS). Because the time between measurements is fixed and relatively long, typically 10s, these radars are limited in their ability to rapidly detect maneuvers and maintain track in the presence of other potential targets, especially crossing ones.
- Dish radars are usually dedicated to track a single target, and scan the antenna mechanically to maintain their narrow beam on the target. They generate measurements at a high rate. Some such radars track multiple targets when they are in the same radar beam position.
- Multi-function phased-array radars track multiple targets by directing the beam electronically to each target. Measurement rates on each target are varied as needed, and many phased-array radars select waveforms having the needed energy and resolution for each target.

Post processing of a series of target measurements can be used to estimate the target path after the fact, using smoothing techniques such as least-mean-square error fit. Such processing is useful for analysis, for example, of tests.

In contrast, real-time target tracking, the focus of this section, maintains estimates of target parameters continuously, so that a user can respond with timely action, such as flight direction or engagement. Such tracking can employ:

- Batch processing, where the measurements are divided into groups, and each group is processed to estimate target parameters.
- Recursive processing, where each new measurement is processed with the results from previous measurements to update the target parameters.

Tracking filters. Radar tracking filters compare radar measurement data over a time interval with a model of target motion. A least-square error or a maximum likelihood estimator is often used to estimate the target parameters [25].

- The α-β filter uses a model of target position, α, and velocity, β. The fixed filter parameters are chosen based on the expected range of target speed (velocity) and maneuverability (acceleration). They are designed as a compromise between smoothing to reduce random measurement errors and responsiveness to target maneuvers. The filter can be implemented as a batch processor, or a sliding-window processor, where the estimate is based on the most recent n measurements.

- The α-β-γ filter adds γ to model target acceleration. This can reduce the dynamic lag of the β filter caused by target maneuver, but the error for non-maneuvering targets may increase. Implementation is similar to that of α-β filters.

- The Kalman filter is a recursive filter that extends the α-β or α-β-γ filter by varying the filter parameters with time. The parameters are based on the expected target motion parameters and the characteristics of the radar measurements, and are updated with each measurement based on the history of past measurements. Kalman filters minimize the mean-square error when the measurements have Gaussian probability distributions. While they are more complex than the fixed-parameter filters, they are often used due to their capabilities for handling missing data, changes in measurement noise, and variations in target dynamics.

Most filters operate in Cartesian (x, y, z), coordinates, while radar measurements are in polar (range, angle, range rate), coordinates. While coordinate transformations are straight forward, they produce non-linearity that can degrade filter performance.

Track association. To maintain track integrity, new measurements must be correctly associated with existing tracks. This becomes increasingly difficult with high target density and many measurements from false detections. Track association techniques in increasing order of complexity include [26]:

- Nearest neighbor. With this technique, a new measurement location is assigned to the nearest predicted target position, taking into account the accuracy of the measurement and the predicted position. This is simple to implement, but can lead to association errors when the accuracy of target positions and the measurement make more than one association possible.

- Global nearest neighbor. Here, the sum of the distances between measurements and assigned targets is minimized taking into account the accuracy of the measurement and the predicted position for each association. This can be optimally done using a Munkres algorithm, but other, less computationally-intense, algorithms are often used.

- Probabilistic association. In this technique, tracks are updated with data from all nearby measurements, weighted by the estimated probability of being the correct association. This is most effective when many of the nearby measurements are from false targets due to noise.

- Multiple hypotheses. In this approach several possible tracks are formed by updating nearby tracks using each new measurement. The resulting tracks are evaluated and only the most probable are retained.

The choice of association algorithm depends on the density of new measurements due to closely spaced targets and to noise detections, on the tracking accuracy, and on the accuracy with which measurements are made.

Tracking rate. In some radars the track update rate is fixed. For example, TWS using rotating surveillance radars. In other radars, such as multi-function phased-array radars, the track rate on each target may be varied, within limitations of power and timeline availability. Factors that affect track-rate selection include:

- Maintain track. The uncertainty in the predicted target position at the time of the next measurement should be smaller than the window of the observation. Usually this window is defined by the radar beamwidths. To assure a high probability of track update, the time between measurements should be small enough that the predicted position error due to measurement accuracy and target maneuver, (typically $3\text{-}\sigma$), are less than the distance corresponding to half the beamwidth, $(R\theta/2)$, in both angular coordinates.

- Track accuracy. The radar tracking error is approximately proportional to the inverse square root of the track rate. Thus increasing track rate can improve tracking accuracy, assuming the same observation S/N.

- Track association. A key factor in associating new measurements with existing tracks is the accuracy of the predicted target position at the time of a new measurement. Increasing the tracking rate improves the predicted position accuracy because of increased track accuracy and reduced prediction time.

4 | Radar Environment

4.1 Atmospheric Losses

The atmosphere, also called the troposphere, produces losses in radar signal propagation due to atmospheric attenuation, and to beam spreading.

Atmospheric attenuation. This loss is caused by molecular absorption by oxygen and water vapor in the atmosphere. The loss increases gradually with frequency in the microwave-frequency region, and has resonant peaks at 22.3 GHz due to water vapor and at 60 GHz due to oxygen. The attenuation decreases with altitude, and usually can be neglected above 10 km. For surface radars, the attenuation decreases with increasing elevation angle, and can usually be neglected at elevation angles above 10 degrees.

Atmospheric loss increases exponentially with path length, l, and can be characterized by a two-way loss in dB per km, a_A (Fig. 4.1). For paths where a_A remains constant, the total loss, L_A, in dB is:

$$L_A(\text{in dB}) = a_A \, l \tag{4.1}$$

The loss power ratio is:

$$L_A \, (\text{power ratio}) = 10^{\frac{a_A l}{10}} \tag{4.2}$$

For signal paths where a_A varies, the attenuation is found by integrating along the signal path. Values for surface radars are given as functions of range and elevation angle for UHF, L, S, C, and X bands in Figs. 4.2 through 4.6 [27].

Lens loss. At low elevation angles the difference in atmospheric refraction (Sec. 4.3), at the top and bottom of the beam increases the elevation beamwidth, reducing the gain. This is characterized as the lens loss. The loss is independent of frequency, increases with range, and decreases with increasing elevation angle. It usually can be neglected at elevation angles greater than 5 degrees. Values of lens loss for surface radars are given as a function of range and elevation angle in Fig. 4.7 [28, Ch, 15]. Values for other radar altitudes can be found in [28, Ch, 15].

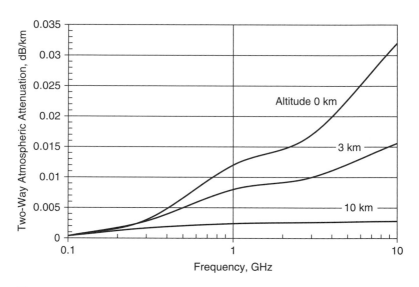

Fig. 4.1 Atmospheric attenuation coefficient, a_A, vs. frequency and elevation angle (made using the custom radar functions in [5])

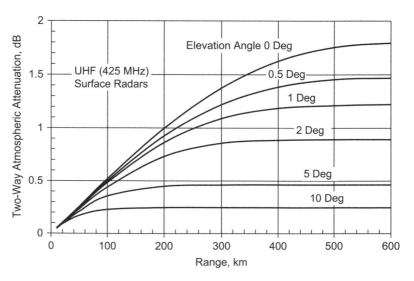

Fig. 4.2 Atmospheric attenuation at UHF vs. range and elevation angle for surface radars (made using the custom radar functions in [5])

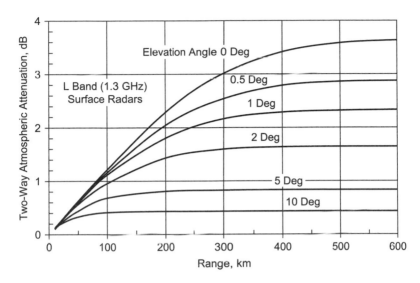

Fig. 4.3 Atmospheric attenuation at L band vs. range and elevation angle for surface radars (made using the custom radar functions in [5])

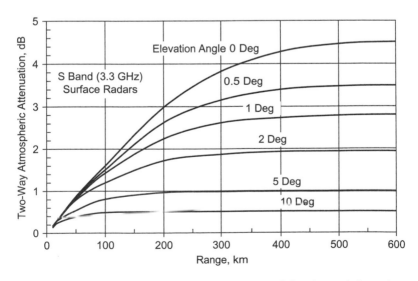

Fig. 4.4 Atmospheric attenuation at S band vs. range and elevation angle for surface radars (made using the custom radar functions in [5])

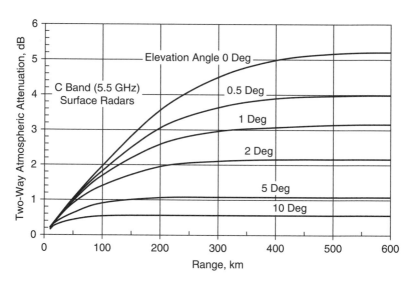

Fig. 4.5 Atmospheric attenuation at C band vs. range and elevation angle for surface radars (made using the custom radar functions in [5])

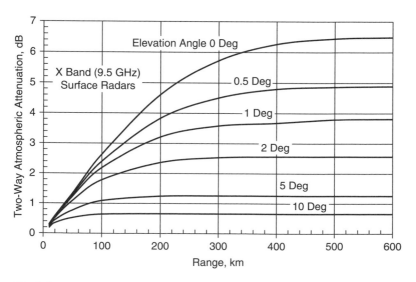

Fig. 4.6 Atmospheric attenuation at X band vs. range and elevation angle for surface radars (made using the custom radar functions in [5])

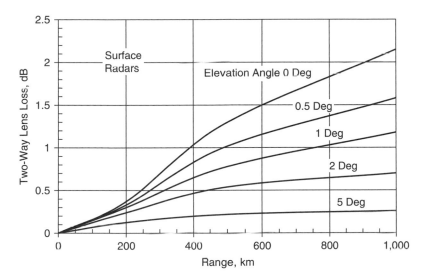

Fig. 4.7 Lens loss vs. range and elevation angle for surface radars (made using the custom radar functions in [5])

The lens loss from Fig. 4.7 should be added to the atmospheric attenuation from Figs. 4.2 through 4.6 to get the total atmospheric losses.

4.2 Rain Loss

Rain causes signal attenuation for signal paths that pass through the rainfall. The attenuation increases significantly with frequency, and can usually be neglected at frequencies below 1 GHz. Rain loss increases exponentially with path length, l, and can be characterized by a two-way loss in dB per km, a_R (Fig. 4.8), [29, Ch. 1]. For paths where a_R remains constant, the total loss, L_R, in dB is:

$$L_R(\text{in dB}) = a_R l \tag{4.3}$$

The loss power ratio is:

$$L_R \ (\text{power ratio}) = 10^{\frac{a_R l}{10}} \tag{4.4}$$

The signal attenuation is found by integrating the attenuation along the signal path. Since only liquid rain produces significant attenuation, only the portion of signal paths below the zero-degree isotherm, (about 3 km altitude at mid latitudes), will experience rain loss. Also, rainfall is not uniform over extended areas. High rainfall rates are usually confined to relatively small areas, typically 10 km or less. Thus, Eqs. 4.3 and 4.4

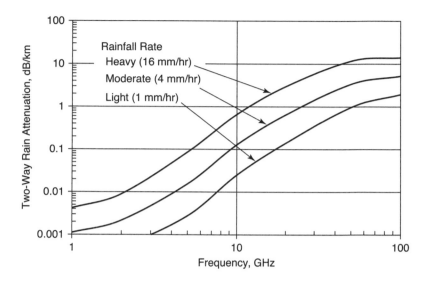

Fig. 4.8 Rain attenuation coefficient, a_R, vs. frequency (made using the custom radar functions in [5])

must be used with care. A model for predicting probable path loss from rainfall was developed by Crane [29, Ch. 4].

4.3 Atmospheric Refraction

Refraction or bending of the radar-signal propagation path in the atmosphere (also called the troposphere), is caused by small variations of the propagation velocity with altitude due to variations in pressure, temperature, and water vapor content. The refractive index, n, is the ratio of the propagation velocity to that in a vacuum. It normally decreases with increasing altitude, causing a downward bending and lengthening of the propagation path.

The atmospheric refractivity, N, is:

$$N = 10^6 (n - 1) \qquad (4.5)$$

It normally decreases approximately exponentially with increasing altitude. The refractivity at the earth's surface has a standard value of 313, and can vary over a range of about ±10%.

Measurement errors. Atmospheric refraction produces errors in radar measurements of range and elevation angle. These errors decrease with increasing elevation angle, and are independent of frequency for frequencies below about 20 GHz (Figs. 4.9 and 4.10), [10, App. D]).

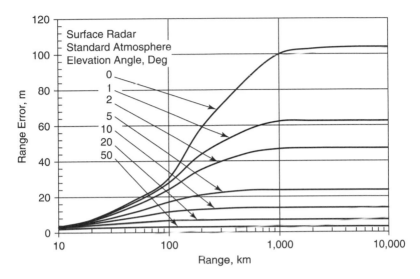

Fig. 4.9   Range-measurement error vs. range and elevation angle for a surface radar and standard atmosphere (made using the custom radar functions in [5])

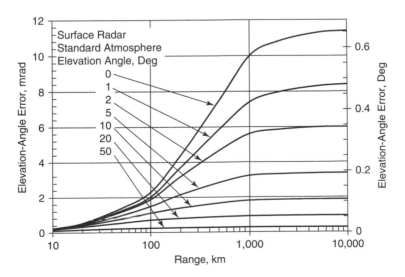

Fig. 4.10   Elevation-angle-measurement error vs. range and elevation angle for a surface radar and standard atmosphere (made using the custom radar functions in [5])

The atmospheric measurement errors can be corrected to an accuracy of about 15% of their magnitude by using data for a standard atmosphere, such as in Figs. 4.9 and 4.10. If the refractivity at the radar is measured, the accuracy of the correction improves to about 5% [6, Ch. 6]. These errors normally remain fixed during an observation, and are treated as bias errors when determining overall measurement accuracy (Sec. 3.5).

4/3 earth model. Atmospheric refraction paths are often modeled by assuming an earth radius that is 4/3 times the actual radius of 6,371 km, giving a value of 8,495 km. With this model, propagation paths within the atmosphere are approximately straight lines. This approximation is accurate for altitudes below 4 km, and is often used for altitudes up to 10 km with small error [5, Ch. 9]

Anomalous propagation (ducting). Atmospheric conditions can occur that produce refractivity above the earth's surface that decreases much more rapidly than in a standard atmosphere. This occurs most frequently over tropical ocean areas. The effect increases the refraction for surface radars, and can produce propagation around the earth's curvature, a phenomenon called surface ducting. This can allow detection of surface and low-altitude targets at ranges much greater than normally, and can also lead to unexpected second- and multiple-time-around returns (Sec. 1.3).

4.4   Terrain Masking and Multipath

Radar horizon range. Blocking of the radar line-of-sight (LOS), by terrain or the sea surface can limit the observation of low-altitude targets by surface or low-altitude radars. The radar horizon range, (the range from the radar to the point where the LOS is tangent to the earth or sea surface), is shown versus radar altitude above a smooth earth in Fig. 4.11.

For radar altitudes less than about 10 km, the horizon range, R_H, can be found using the 4/3 earth model (Sec. 4.3):

$$R_H = \left(h_R^2 + 2\, r_E\, h_R \right)^{1/2} \tag{4.6}$$

where h_R is the radar altitude above a smooth earth and r_E is the 4/3 earth radius, 8,495 km. The horizon range from the target to the LOS tangent point can also be found from Fig. 4.11 or Eq. 4.6, using the target altitude above a smooth earth.

Maximum range limitation. The sum of the radar horizon range and the target horizon range is the maximum range at which the target can be observed, given the radar and target altitudes and a smooth earth.

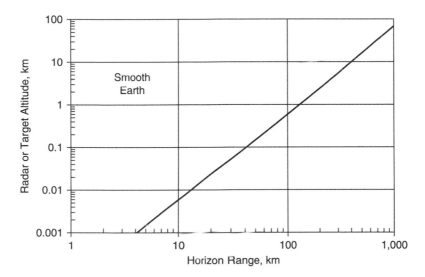

Fig. 4.11   Radar or target altitude vs. horizon range for a smooth earth

In many cases, radars are located at elevated sites to increase their horizon range. Conversely, hilly or mountainous terrain can reduce the radar or target horizon range. Also, it is often desirable to view targets at least a few degrees above the local horizon to avoid multipath propagation (see below), and to reduce atmospheric loss and refraction (Sec. 4.1 and 4.3). This will reduce the range at which low-altitude targets can be observed below that determined by the horizon ranges.

Multipath geometry. When the radar beam illuminates the terrain or sea surface between the radar and the target, as well as the target, a second, reflected, signal path between the radar and target is created. When the grazing angle (the angle between the earth's surface and the signal path), is small, the range difference between the direct signal path and the reflected signal path, δR, is small, and given approximately by:

$$\delta R \approx \frac{2 h_R h_T}{R} \tag{4.7}$$

where h_R is the radar altitude, h_T is the target altitude, and R is the target direct-path range.

When δR is smaller than the radar range resolution, ΔR, the direct and reflected signals add coherently at the radar receiver, either increasing or decreasing the received signal, depending on their relative phase. This effect is called multipath propagation. It can occur with:

- Rotating search radars that employ broad elevation-angle fan beams (Sec. 2.1).
- Pencil-beam antennas (Sec. 2.1), observing targets near the radar horizon.
- Airborne radars observing low-altitude targets with low signal-path grazing angles.
- Large structures (buildings, dams, etc.), that are illuminated by the radar beam and create a second signal path.

Multipath effects. With a flat, perfectly-reflecting surface, the received signal varies between zero and 16 times the signal level from the direct path alone. The resulting range for a given S/N varies between zero and twice the range when only the direct path is present due to the R^4 dependence on S/N. When the reflecting surface is not perfectly reflecting, or it is curved, the received signal peaks and nulls are less pronounced.

When the grazing angle is near zero, the received signal level and range are near their minimum, (zero for flat, perfectly-reflecting surfaces). As the elevation angle increases, a series of lobes are produced, separated in elevation angle by $\Delta\phi$, given by:

$$\Delta\phi = \frac{\lambda}{2\,h_R} \tag{4.8}$$

Multipath conditions can produce significant angle-measurement errors at low elevation angles [10, Ch. 5]

- For smooth reflecting surfaces, elevation-angle measurement errors are about half the elevation beamwidth for target elevations less than 0.8 of the elevation beamwidth. Smooth reflecting surfaces produce no azimuth-measurement errors.
- For rough reflecting surfaces and elevation angles less than the elevation beamwidth, elevation-angle measurement errors are typically 0.1 of the elevation beamwidth, and azimuth measurement errors are 0.1 to 0.2 of the azimuth beamwidth.

Measurement errors from multipath fluctuate slowly as the observation geometry changes, and should be treated as bias errors when combining them with other measurement errors (Sec. 3.5).

4.5 Radar Clutter

Radar returns from terrain or sea surface, or from rain, that interfere with the desired target signal are termed clutter. The interference level can be characterized by the ratio of the received signal power, S, to the clutter power, C. This signal-to-clutter ratio, S/C, depends on the target RCS, the amount of clutter illuminated, the clutter reflectivity and the effect of any clutter reduction techniques (Table 4.1). It is independent of radar sensitivity.

When the clutter signal has Gaussian probability distribution and is random from pulse-to-pulse, S/C can be combined with S/N:

$$\frac{S}{C+N} = \frac{1}{\frac{1}{S/C} + \frac{1}{S/N}} \tag{4.9}$$

This is often the case when many comparable scatterers are in the clutter resolution cell. Then, $S/(C+N)$ can be used in place of S/N in calculating radar detection and measurement performance (Sec. 3.3 and 3.5).

Surface clutter geometry. In most cases, the primary surface clutter that interferes with the target is from the same range-angle resolution cell as the target. (Sidelobe clutter can be significant for airborne radars using pulse-Doppler waveforms, and is discussed in Sec. 5.3.) The surface area of the clutter resolution cell, A_C, is usually given by:

$$A_C = \frac{R\,\theta_A\,\Delta R}{\cos\gamma} \qquad \left[\frac{\pi\,R\,\theta_E}{4\,\Delta R} \geq \tan\gamma\right] \tag{4.10}$$

TABLE 4.1 Characteristics of Surface and Rain Clutter

Clutter Source	Terrain or Sea	Rain
Clutter region	$A_C = \dfrac{R\,\theta_A\,\Delta R}{\cos\gamma}$	$V_C = \dfrac{\pi\,R^2\,\theta_A\,\theta_E\,\Delta R}{4}$
Clutter reflectivity	σ^0(m^2 RCS/m^2 surface)	η(m^2 RCS/m^3 volume)
Reflectivity factors	Depends on frequency, grazing angle γ, polarization, terrain, sea state	$\eta = \dfrac{6\times10^{-14}\,\text{rain rate (mm/hr)}^{1.6}}{\lambda^4}$
Clutter-reduction techniques (CR)	MTI Pulse-Doppler	Pulse-Doppler Polarization
Signal-to-clutter ratio	$\dfrac{S}{C} = \dfrac{\sigma\,\text{CRL}_{BS}}{\sigma^0\,A_C}\,(L_{BS}\approx1.5)$	$\dfrac{S}{C} = \dfrac{\sigma\,\text{CRL}_{BS}}{\eta\,V_C}\ \ (L_{BS}\approx2.1)$

where R is the target range, θ_A is the azimuth beamwidth, θ_E is the elevation beamwidth, ΔR is the range resolution, and γ is the angle between the radar line-of-sight (LOS), and the surface, called the grazing angle.

For high grazing angles and large range resolution, the range extent of the clutter area may be determined by the elevation beamwidth, θ_E, rather than by the range resolution. Then A_C is:

$$A_C = \frac{\pi R^2 \theta_A \theta_E}{4 \sin \gamma} \qquad \left[\frac{\pi R \theta_E}{4 \Delta R} \leq \tan \gamma \right] \qquad (4.11)$$

The clutter reflectivity is characterized by a parameter σ^0, which is dimensionless and is equal to the square meters of clutter RCS per square meter of surface area. The parameter σ^0 is usually significantly less than unity, and is often expressed by a negative dB value. (A larger reflectivity leads to a smaller negative dB value.) The clutter RCS in the resolution cell is given by:

$$\sigma_C = \frac{\sigma^0 A_C}{L_{BS}} \qquad (4.12)$$

Where L_{BS} is a beam-shape loss for the clutter, equal to about 1.5 (1.6 dB). The signal-to-clutter ratio is:

$$\frac{S}{C} = \frac{\sigma \, CRL_{BS}}{\sigma^0 A_C} \qquad (4.13)$$

where CR is the clutter cancellation ratio, the improvement in S/C provided by clutter-reduction techniques such as those described later.

Surface reflectivity. The magnitude of the surface reflectivity, σ^0, is determined by several factors [30, 31, and 32]:

- Grazing angle. σ^0 varies approximately with the sine of grazing angle, γ, between grazing angles of a few degrees and about 60 deg. At the lower grazing angles, multipath interference produces lower reflectivity, while at high elevation angles specular returns result in higher values.
- Frequency. Sea clutter reflectivity is approximately proportional to frequency. For terrain, σ^0 generally increases with increasing radar frequency.
- Terrain type. σ^0 is generally larger for rougher terrain and for more pronounced vegetation. Large discrete returns, for example from buildings, produce non-Gaussian probability distributions.

- Sea state. σ^0 increases with sea state.
- Polarization. For smooth terrain and sea surface and low grazing angles, σ^0 is greater for vertical polarization than for horizontal polarization. For rough terrain and at high grazing angles, the difference in reflectivity between polarizations is small.

Representative plots of terrain and sea clutter reflectivity vs. grazing angle are shown in Figs. 4.12 and 4.13.

Surface clutter usually has a small velocity spread due to the motion of the clutter scatterers, producing a small Doppler-frequency spread for stationary radars:

- For terrain, the velocity spread ranges from near zero for rocky terrain, to about 0.33 m/s for wind-blown trees.
- For sea clutter, the velocity spread is approximately 0.125 times the wind velocity.
- Rotating search radars produce a velocity spread component equal to the velocity of the antenna edge.

Rain clutter. In most cases, the primary rain clutter that interferes with the target is from the same range-angle resolution cell as the target.

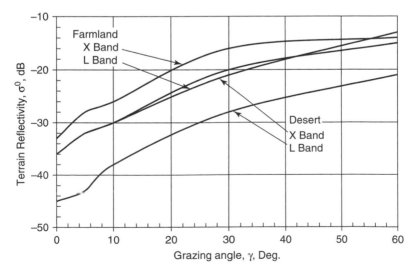

Fig. 4.12 Average terrain clutter reflectivity vs. grazing angle for L and X bands (data from [30])

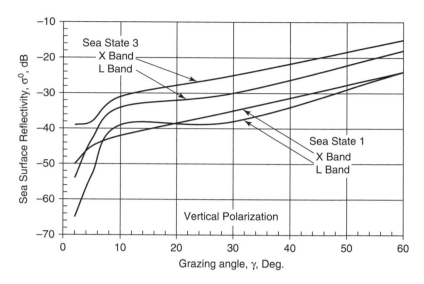

Fig. 4.13 Average sea clutter reflectivity vs. grazing angle for L and X bands (data from [30])

The volume of the clutter resolution cell, V_C, is:

$$V_C = \frac{\pi R^2 \theta_A \theta_E \Delta R}{4} \qquad (4.14)$$

where R is the target range, θ_A and θ_E are the azimuth and elevation beamwidths, ΔR is the range resolution. Equation 4.14 assumes that rain fills the radar resolution cell. When this is not so, the actual volume of rain in the resolution cell should be used.

Rain clutter reflectivity is characterized by a volume reflectivity parameter, η, which has the dimension m^{-1}, (square meters of clutter per cubic meter of rain), and is given by:

$$\eta = \frac{6 \times 10^{-14} r^{1.6}}{\lambda^4} \qquad (4.15)$$

where r is the rainfall rate in mm/hr (Fig. 4.14). Equation 4.15 shows that rain clutter increases with rainfall rate and with the fourth power of frequency.

The clutter RCS in the resolution cell is given by:

$$\sigma_C = \frac{\eta V_C}{L_{BS}} \qquad (4.16)$$

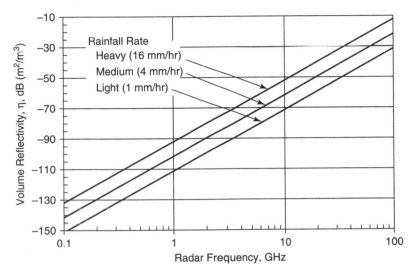

Fig. 4.14 Rain volume reflectivity vs. frequency and rainfall rate

where L_{BS} is a beam-shape loss for the clutter, equal to about 2.1 (3.2 dB). The signal-to-clutter ratio is:

$$\frac{S}{C} = \frac{\sigma \, \text{CRL}_{BS}}{\eta \, V_C} \tag{4.17}$$

The S/C in rain (Eq. 4.17) varies inversely with η, and decreases with rainfall rate and radar frequency.

The velocity of rain clutter is generally that of the wind. Velocity spread is of the order of 2 to 4 m/s.

Clutter reduction. Since clutter results mainly from scatterers in the radar resolution cell, it is minimized by using small radar angle resolution (narrow beamwidth), and small range resolution (wide signal bandwidth). Other techniques for clutter reduction include:

- Moving-target indication (MTI), used by surface radars to cancel clutter (Sec. 5.2).
- Pulse-Doppler processing, used by airborne and space-based radars, as well as some surface radars to avoid and reject clutter (Sec. 5.3).
- Radars using circular polarization can reject rain clutter by receiving the same sense circular polarization transmitted. This rejects spherical raindrops, while having only a small effect on the signal from most complex targets (Sec. 3.1).

The ability of these techniques to reduce clutter depends on the clutter characteristics, radar stability and dynamic range, and details of clutter canceller design and associated signal processing. Typical cancellation ratios (CR) are in the 20–40 dB range.

4.6 Ionospheric Effects

The ionosphere comprises layers of ionized electrons at altitudes between about 55 and 1,000 km. Radar propagation paths that pass through the ionosphere can be affected. These paths include surface and airborne radars observing objects in space, and space-based radars observing terrestrial and airborne targets. The ionospheric effects on radar propagation vary inversely with various powers of radar frequency and are rarely significant at frequencies above about 1 GHz [33].

Ionospheric characteristics. Ionospheric effects on propagation depend on the integrated electron density along the signal path. For signal paths through the ionosphere, this decreases with increasing elevation angle. For paths ending in the ionosphere, it increases within range.

The electron density in the ionosphere is highly variable, and depends on:

- Solar radiation. It is significantly higher in the daytime than at night.
- Latitude. It is greatest at about 20 degrees latitude and in polar regions.
- Sunspot activity. It is higher in periods of increased sunspot activity.

This variability makes it difficult to predict the magnitude of ionospheric effects on radar propagation, and to correct for them.

Ionospheric attenuation. The attenuation from the ionosphere (in dB), varies inversely with the square of frequency. It is usually significant only at frequencies below about 300 MHz. The normal daytime two-way attenuation for signal paths passing through the ionosphere is shown vs. elevation angle for VHF and UHF frequencies in Fig. 4.15. At nighttime, the ionospheric attenuation is less than 0.1 dB for all elevation angles [33].

The non-uniform electron density in the ionosphere can produce random fluctuations in signal amplitude and phase in the VHF and UHF bands, and sometimes at L band. While these do not affect the average signal power, they can affect detection performance, especially when coherent integration is used (Sec. 3.3).

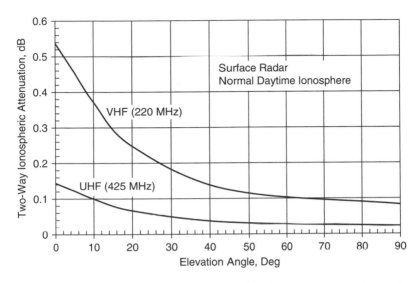

Fig. 4.15 Ionospheric attenuation at VHF and UHF frequencies vs. elevation angle for normal daytime ionospheric conditions (data from [33])

Ionospheric refraction. The refractive index in the ionosphere is less than unity, causing the signal path for a terrestrial radar to bend downward as it enters the ionosphere, and then upward as it leaves the ionosphere, so that is parallel with, but offset from, the signal path entering the ionosphere. This produces elevation-angle and range errors, which vary inversely with the square of frequency.

These errors are shown for a surface-based radar with a normal daytime ionosphere vs. range and elevation angle for VHF and UHF frequencies in Figs. 4.16 through 4.19 [33]. Nighttime errors are about one third of those plotted in the figures, while daytime errors during periods of ionosphere disturbance may be as much as three times those plotted. This wide range makes it difficult to correct for ionospheric refraction errors, unless the ionospheric conditions are measured.

Frequency dispersion. The variation of refractive index with frequency produces frequency dispersion of signals passing through the ionosphere. This limits the maximum signal bandwidth that can be supported by such signal paths. The maximum bandwidth varies as the 1.5 power of frequency.

Approximate values of maximum signal bandwidth for signals passing through the ionosphere are given vs. frequency and elevation angle

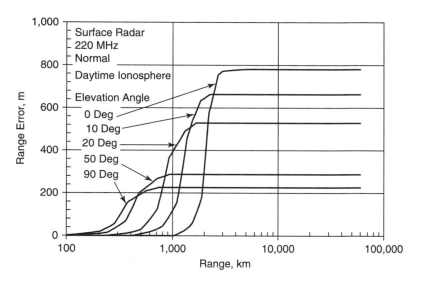

Fig. 4.16 Range errors at 220 MHz for normal daytime ionospheric conditions vs. range and elevation angle (made using the custom radar functions in [5])

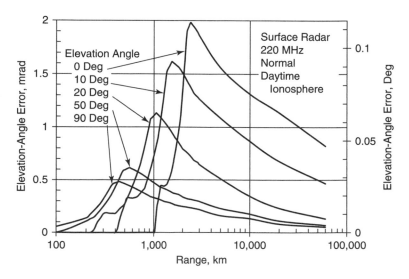

Fig. 4.17 Elevation-Angle errors at 220 MHz for normal daytime ionospheric conditions vs. range and elevation angle (made using the custom radar functions in [5])

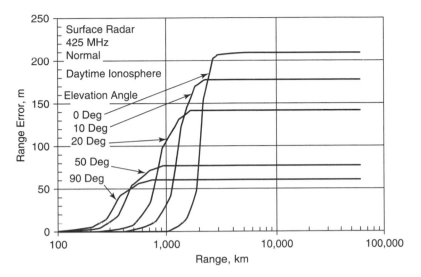

Fig. 4.18 Range errors at 425 MHz for normal daytime ionospheric conditions vs. range and elevation angle (made using the custom radar functions in [5])

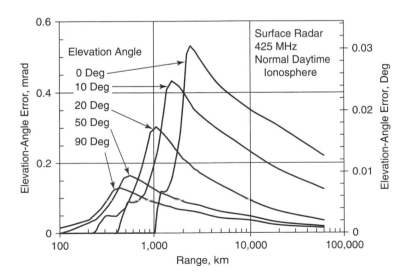

Fig. 4.19 Elevation-Angle errors at 425 MHz for normal daytime ionospheric conditions vs. range and elevation angle (made using the custom radar functions in [5])

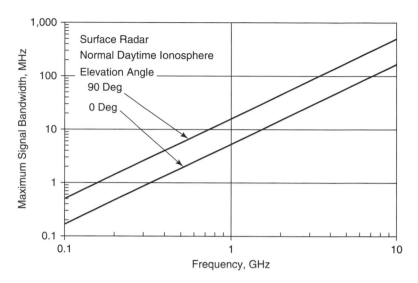

Fig. 4.20 Approximate maximum bandwidth for signals traversing the ionosphere vs. frequency and elevation angle for normal daytime ionosphere (data from [33])

for a normal daytime ionosphere in Fig. 4.20. Compensation for the estimated dispersion may allow larger bandwidths to be used. Nighttime bandwidths are significantly larger than in daytime.

Polarization rotation. The polarization of linearly-polarized signals rotates when passing through the ionosphere. The magnitude of this rotation is inversely proportional the square of frequency. At frequencies below about 2 GHz, the polarization of the received signal may not match that of the transmitted signal [33]. This could produce a significant signal loss unless dual linear polarizations are received. This condition may be mitigated by using circular polarization, which is not affected by ionospheric rotation.

4.7   Electronic Countermeasures (ECM)

Radars may be subject to deliberate electronic interference measures, termed electronic countermeasures (ECM). These can be categorized as (Table 4.2):

- Masking of radar targets, often using noise jamming or radar chaff.
- Confusion, by producing many false targets, often using pulse jamming or physical objects.

TABLE 4.2 Common Radar Countermeasures and Counter-Countermeasures (source: [5, Ch 10])

Category	Countermeasure	Counter-Countermeasures
Masking	Mainlobe jammer	Frequency agility
		Burnthrough
		Passive tracking
	Sidelobe jammer	Low sidelobes
		Frequency agility
		Burnthrough
		Sidelobe cancellers
	Volume chaff	MTI and pulse-Doppler
		Range resolution
		Velocity resolution
Confusion	Pulsed jammer	Sidelobe blanker
		Frequency agility
		Tracking
	Traffic decoys	Range resolution
		Bulk filtering
		Tracking
	Debris	Bulk filtering
		Tracking
Deception	Repeater jammer	Frequency agility
		PRF agility
		Signal processing
	Track-breaking jammer	Signal processing
		Tracking
	Spot chaff	MTI and pulse Doppler
		Radar measurements
		Tracking
	Decoy	Radar measurements
		Tracking

- Deception by creating false targets, either electronically or using decoy objects, or by breaking track on actual targets.

Radar designers and operators employ electronic counter-countermeasures (ECCM), to reduce the impacts of ECM on radar performance. These can include radar design features, operating modes, and dedicated ECCM techniques.

Common ECM techniques and potential ECCM responses are summarized below.

Noise jamming. Noise jammers attempt to mask radar targets by increasing the level of the noise competing with the target signal. Noise-jammer effectiveness is characterized by a signal-to-jammer power ratio, S/J. When the jammer noise has Gaussian probability density, S/J can be combined with S/N to give an overall signal-to-interference ratio:

$$\frac{S}{J+N} = \frac{1}{\frac{1}{S/J} + \frac{1}{S/N}} \tag{4.18}$$

Then $S/(J+N)$ can be used in place of S/N in calculating radar detection and measurement performance (Sec. 3.3 and 3.5). Radars often use constant-false-alarm-rate, (CFAR), techniques to automatically raise the detection threshold in the presence of jamming to maintain the false-alarm rate (Sec. 3.3).

A noise jammer can be described by its noise bandwidth, B_J, and its effective radiated power (ERP), which is the power density radiated in the direction of the radar to be jammed. The ERP is given by:

$$\text{ERP} = \frac{P_J \, G_J}{L_J} \tag{4.19}$$

where P_J is the jammer RF power, G_J is the jammer antenna gain in the direction of the radar, and L_J accounts for any jammer losses. ERP is often specified in dB relative to a Watt, dBW (Sec. 6.3).

Only the jammer noise power within the radar signal bandwidth, B, is effective in masking the target. However, if the jammer does not know the exact radar signal frequency, or if the radar changes its frequency from pulse to pulse (frequency agility), the jammer may have to jam a much wider bandwidth, B_J, to assure that the radar band is jammed. This is called barrage jamming.

Equation 4.19 assumes that the jammer polarization is aligned with that of the radar to be jammed. If the jammer does not know the radar polarization or can not control its orientation, two jammers with independent noise sources and orthogonal polarizations must be used. The ERP from Eq. 4.19 is then referred to as the ERP per polarization.

Jammer performance and radar ECCMs depend on whether the jammer is in the main radar beam or in a sidelobe region:

- Mainlobe jammers (MLJ). These signals enter the radar through the radar main radar beam. The jammers may be located on the target itself and called self-screening jammers (SSJ), or they may be located on a vehicle near the target and called escort jammers (ESJ). The S/J for a MLJ is given by:

$$\frac{S}{J} = \frac{P_P \, G_T \, \sigma \, \text{PC} \, B_J}{4\pi \, R^2 \, BL_T \, \text{ERP}} \tag{4.20}$$

where L_T is the radar transmit losses, and $B_J \geq B$. The S/J is determined by the signal reflected by the target and the jammer ERP, and is independent of receiver parameters. Like S/N, S/J can be increased by using pulse integration (Sec. 3.2). The range at which a target may be observed with a desired S/J is called the burnthrough range:

$$R = \left[\frac{P_P \, G_T \, \sigma \, \text{PC} \, B_J}{4\pi \, (S/J) \, BL_J \, \text{ERP}} \right]^{1/2} \tag{4.21}$$

Since MLJs are at or near the target, passively tracking them can reveal the target location.

• Sidelobe jammers (SLJ). These signals enter the radar through the radar sidelobes. The jammers are often deployed at long ranges, outside the coverage of hostile weapons. Then they are called stand-off jammers (SOJ). The S/J for a SLJ is given by:

$$\frac{S}{J} = \frac{P_P \, G_T \, \sigma \, \text{PC} \, R_J^2 \, B_J}{4\pi \, R^4 \, B \, \text{SL}L_T \, \text{ERP}} \tag{4.22}$$

where R_J is the jammer range and SL is the radar sidelobe level at the jammer angle (a value less than unity, often represented by a negative dB value). Sidelobe jammers usually have higher power and gain than mainlobe jammers to overcome both the radar sidelobes and any increase in range. This requires that the jammer antenna be directed toward the targeted radar. The burnthrough range for SLJs is:

$$R = \left[\frac{P_P \, G_T \, \sigma \, \text{PC} \, R_J^2 \, B_J}{4\pi \, (S/J) \, B \, \text{SL} \, L_J \, \text{ERP}} \right]^{1/4} \tag{4.23}$$

Sidelobe cancellers (SLC) reduce the sidelobe level at the jammer angle. These use an auxiliary antenna to sense the jamming signal, adjust its amplitude and phase, and subtract it coherently from the main radar receive channel. Multiple SLCs can be used to cancel multiple jammers. Phased array radars may calculate complex element weighting that minimizes the sidelobe levels at jammer angles. SLCs can reduce sidelobes by 20 dB or more.

Pulse jamming. Pulsed jammers produce false radar signals that confuse or deceive the radar operator or signal processor. Common techniques include:

- Sidelobe confusion jammers can produce large numbers of false targets in the radar sidelobe region and require radar and processing resources to confirm or reject them. Radars can reject sidelobe pulsed jammers by using sidelobe blankers (SLB). These use an auxiliary antenna to sense the timing of pulses entering the sidelobes, and blank the main receiver during these periods. These differ from the SLCs discussed above in that the radar is blanked during the pulse-arrival times. They can not be used with continuous jammers or the radar would be inoperative.
- Swept-spot jammers can create false targets as the jammer sweeps through the radar signal band. By using very-rapid sweep rates they can produce an effect similar to broadband noise and mask targets over the radar band.
- Repeater jammers transmit replicas of the radar signal to create false targets. These may be synchronized with the radar pulse transmissions to generate stationary targets or those on credible paths. If the radar frequency changes, the repeater frequency must follow those changes. This limits the false-target positions to ranges greater than that of the jammer, which is often located on a real target.
- Track-breaking jammers are usually located on a target that is being tracked by the radar. They emit sophisticated signals timed and phased to move the range or angle tracking gates off the real target so that track is lost.

Sidelobe blanking, frequency agility and PRF changes reduce the effectiveness of many pulsed jamming techniques. Radar signal and data processors are sized to handle the remaining target load.

Volume chaff. Radar volume chaff consists of a large number of small scatterers located around a target of interest. The reflections from the chaff are intended to mask the radar signal return from the target. Chaff may be deployed in a cloud around a target, or in a corridor to mask targets that pass through.

Chaff effectiveness is characterized by a signal-to-chaff RCS ratio, S/C. When the chaff return has Gaussian probability density, S/C can be combined with S/N to give an overall signal-to-interference ratio:

$$\frac{S}{C+N} = \frac{1}{\frac{1}{S/C} + \frac{1}{S/N}} \tag{4.24}$$

Then $S/(C+N)$ can be used in place of S/N in calculating radar detection and measurement performance (Sec. 3.3 and 3.5).

Dipoles that are resonant at the radar frequency are often used for chaff. They have a length of $\lambda/2$ and an RCS average over all angles of $0.15\lambda^2$ [20]. The total RCS, σ_C, for n_C dipoles is:

$$\sigma_C = 0.15\, n_C\, \lambda^2 \tag{4.25}$$

The total chaff RCS can be empirically related to the chaff weight. For common chaff technology [6, Ch. 3]

$$\sigma_C \approx 22,000\, \lambda\, W_C \tag{4.26}$$

where W_C is the total chaff weight in kg. Note that the overhead weight for deploying chaff often equals the chaff weight itself.

Dipole chaff covers a signal bandwidth of about 10%. For larger bandwidths, or to cover multiple radar bands, multiple chaff lengths are used.

In most cases, chaff that interferes with the target is in the same range-angle resolution cell as the target. The volume of the chaff resolution cell, V_C, is:

$$V_C = \frac{\pi\, R^2\, \theta_A\, \theta_E\, \Delta R}{4} \tag{4.27}$$

where R is the target range, θ_A and θ_E are the azimuth and elevation beamwidths, ΔR is the range resolution. Equation 4.27 assumes that chaff fills the radar resolution cell. When this is not so, the actual volume of chaff in the resolution cell should be used.

If the chaff is uniformly distributed over a volume, V_T, then the chaff RCS in the resolution cell that interferes with the target, C, is:

$$C = \frac{\sigma_C\, V_C}{V_T\, L_{BS}} \tag{4.28}$$

where L_{BS} is a beam-shape loss for the chaff, equal to about 2.1, (3.2 dB). The signal-to-chaff ratio is:

$$\frac{S}{C} = \frac{\sigma \, V_T \, L_{BS} \, CR}{\sigma_C \, V_C} \tag{4.29}$$

where CR is a chaff cancellation ratio. Since chaff is usually not uniformly distributed, Equations 4.28 and 4.29 should be considered as approximate.

Chaff deployed in the atmosphere quickly slows to the air speed, while airborne targets usually have much larger velocity. MTI (Sec. 5.2), or pulse-Doppler processing (Sec. 5.3), can provide chaff cancellation of the order of 20–40 dB. Exoatmospheric chaff has a velocity spread produced by its dispensing mechanism, and waveforms having good radial-velocity resolution (Sec. 5.1), may provide useful chaff cancellation.

Objects. Physical objects may be used to confuse or deceive radar operations. These can include:

- Traffic decoys, small objects having observables roughly similar to the targets to be protected. The objective is to overload the radar processor. A radar may use simple thresholds or bulk-filtering techniques such as RCS, or RCS fluctuation rate to reject these.
- Debris from rocket or missile stages and deployment mechanisms may have effects similar to that of deliberate traffic decoys.
- Spot chaff, small clusters of chaff particles, may create false targets. These may be rejected based of radar observables or velocity in the atmosphere.
- Replica decoys are objects designed to match a target's radar observables, and often travel on credible flight paths. Techniques for discriminating between these and actual targets are discussed in Sec. 5.5.

5 | Radar Techniques

5.1 Waveforms

The radar waveform determines, to a large extent, the type and quality of target information that a radar can gather.

Waveform characteristics. The key characteristics of radar waveforms are:

- Energy. The signal-to-noise ratio, (S/N), is directly proportional to the waveform energy, (Sec. 3.2), which affects the target detectability and measurement accuracy, (Secs. 3.3 and 3.5). When the radar peak power, P_P, is constant, the waveform energy, E, is:

$$E = P_P \, \tau \qquad (5.1)$$

  where τ is the waveform duration. Using long waveforms must be consistent with the transmitter capability, (Sec. 2.2), and the minimum-range constraint, (Sec. 1.3).

- Resolution. Waveform resolution determines the radar's capability to separate targets closely spaced in range or radial velocity. Targets must be separable in range, radial velocity, or angle to allow them to be counted and separately observed and tracked. The range resolution, ΔR, is due to the separation of the signal returns in time. For matched-filter signal processing the time resolution, τ_R, is the reciprocal of the signal bandwidth, B:

$$\Delta R = \frac{c \, \tau_R}{2} = \frac{c}{2 \, B} \qquad (5.2)$$

  The radial-velocity resolution, ΔV, is due to the separation of the signal spectra in frequency. For matched-filter processing, the frequency resolution, f_R, is the reciprocal of the waveform duration, τ:

$$\Delta V = \frac{\lambda \, f_R}{2} = \frac{\lambda}{2 \, \tau} \qquad (5.3)$$

- Unwanted-target rejection. Signal returns may result from waveform range or radial-velocity sidelobes, or from ambiguous waveform responses. The waveform must be designed to suppress these, at least in those range-velocity regions that can interfere with desired targets.

The radar ambiguity function [34,35] shows that summation of all waveform responses is a constant. Improving one resolution occurs at the expense of the other resolution, or of sidelobe or ambiguous responses. Thus waveform design and selection involves tradeoffs and matching of waveform characteristics to information needs and to target and interference characteristics. Common waveform types are summarized in Table 5.1, and discussed below [5, Ch. 4, 34, 35].

Time-bandwidth product. The product of the range and radial-velocity resolutions, (Eq. 5.2 and 5.3), is inversely proportional to radar frequency and to the quantity τB, which is called the time-bandwidth product:

$$\Delta R \, \Delta V = \frac{\lambda c}{4\tau B} = \frac{c^2}{4 f \tau B} \tag{5.4}$$

The time-bandwidth product is a measure of the waveform capability to provide both small range resolution and small radial-velocity resolution. It varies from unity for a constant-frequency pulse to greater than 10^8 for some pulse-compression waveforms (Fig. 5.1). When a pulse-compression waveform is compared with a constant-frequency pulse, the range resolution is reduced by τB. Thus the pulse-compression factor, PC, is:

$$\text{PC} = \tau B \tag{5.5}$$

TABLE 5.1 Features of Common Waveform Types

Waveform Type	Pulse-Compression Ratio, ($PC = \tau B$)	Key Limitations
Constant-frequency pulse	1	Can not give good range and velocity resolution
Linear-FM pulse (Chirp)	$>10^4$	Range-velocity coupling
Phase-coded pulse	100–1,000	High sidelobes
Pulse burst	$>10^8$	Range and velocity ambiguities

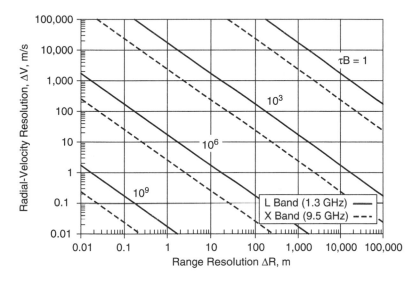

Fig. 5.1 Range and radial-velocity resolution for various time-bandwidth products, τB, and two radar frequencies

For matched-filter signal processing, PC is also equal to the signal power gain from pulse compression (Sec. 3.2).

Constant-frequency pulse. This waveform, also called a continuous-wave, (CW), pulse or simple pulse, consists of a constant amplitude, constant frequency pulse of duration τ. The resolution parameters are:

- Time resolution, $\tau_R = \tau$
- Range resolution, $\Delta R = \dfrac{c\tau}{2}$
- Frequency resolution, $f_R = \dfrac{1}{\tau}$
- Radial-velocity resolution, $\Delta V = \dfrac{\lambda}{2\tau}$
- Time-bandwidth product, $\tau B = 1$

Constant-frequency waveforms are simple to generate and process, but they generally can not provide both good range resolution and good radial-velocity resolution (Fig. 5.1). They are usually used to provide only useful range resolution, with the pulse duration, τ, selected to provide the desired range resolution. In cases where long pulses are used to provide useful radial-velocity resolution, the spectral width of the matched filter will be small. Then multiple matched filters (e.g., a filter bank) are needed to accommodate the expected extent of the target radial velocities.

Linear frequency-modulated, (FM), pulses. These waveforms employ a constant amplitude pulse of duration τ, with a frequency that increases or decreases linearly with time during the pulse over a bandwidth B. They are often called chirp pulses because of the sound of audio pulses with these characteristics. The resolution parameters are:

- Time resolution, $t_R = \dfrac{1}{B}$
- Range resolution, $\Delta R = \dfrac{c}{2B}$
- Frequency resolution, $f_R = \dfrac{1}{\tau}$
- Radial-velocity resolution, $\Delta V = \dfrac{\lambda}{2\tau}$
- Time-bandwidth product $= \tau B$

Linear FM pulses can provide both good range and good radial-velocity resolution with time-bandwidth product, τB, of 10^4 or greater. They are tolerant of wide Doppler-frequency shifts, and can be processed with a single matched filter. Long waveforms may be used to produce high waveform energy while also providing good range resolution.

With linear FM waveforms, the target range response is offset from the true range by an amount proportional to the target radial velocity. This is called range-Doppler coupling. The range offset, R_O, is:

$$R_O = \frac{\tau f V_R}{B} \tag{5.6}$$

The range can be corrected for the offset if the radial velocity is know, for example from the target track, or the measured ranges from frequency modulation in opposite directions, (up and down chirp), can be averaged to obtain the true range.

Phase-coded waveforms. These waveforms consist of a series of n_S subpulses having duration τ_S, each with a specific phase relative to the other subpulses. The total waveform duration is:

$$\tau = n_S \tau_S \tag{5.7}$$

The most common such waveform is the binary-phase coded or phase-reversal waveform, where the relative phases of the subpulses are either

0 or 180 degrees. The resolution parameters are:

- Time resolution, $t_R = \tau_S$
- Range resolution, $\Delta R = \dfrac{c\,\tau_S}{2}$
- Frequency resolution, $f_R = \dfrac{1}{\tau}$
- Radial-velocity resolution, $\Delta V = \dfrac{\lambda}{2\,\tau}$
- Time-bandwidth product $= n_S$

Phase-coded waveforms can provide both good range and good radial-velocity resolution with typical time-bandwidth products, τB, in the 100–1,000 range. However, their sidelobe levels are higher than those of constant-frequency and linear FM waveforms. They are not tolerant of wide Doppler-frequency shifts, and may require multiple matched filters (e.g., a filter bank), to accommodate the expected extent of the target radial velocity. Long waveforms may be used to produce high waveform energy while also providing good range resolution.

Pulse-burst waveforms. These waveforms employ a train of n_S sub-pulses, each with duration τ_S, and spaced in time by τ_P. The total wave-form duration, τ, is:

$$\tau = \tau_S + (n_S - 1)\tau_P \approx n_S\tau_P \tag{5.8}$$

The subpulses may be constant-frequency pulses having bandwidth $B_S = 1/\tau_S$, or they may be compressed pulses having a wider signal bandwidth, B_S. The entire pulse train is processed coherently. The resolution parameters are:

- Time resolution, $t_R = \dfrac{1}{B_S}$
- Range resolution, $\Delta R = \dfrac{c}{2\,B_S}$
- Frequency resolution, $f_R = \dfrac{1}{\tau} \approx \dfrac{1}{n_S\,\tau_P}$
- Radial-velocity resolution, $\Delta V = \dfrac{\lambda}{2\,\tau} \approx \dfrac{\lambda}{2\,n_S\,\tau_P}$
- Time-bandwidth product $= \tau B_S \approx n_S\tau_P B_S$

Pulse-burst waveforms having long bursts and high subpulse bandwidth can have time-bandwidth products, τB, of greater than 10^8, and provide both very-good range resolution and very-good radial velocity resolution.

These waveforms produce ambiguity peaks that are periodic in both range and radial velocity, with spacings:

- Time spacing: τ_P
- Range spacing: $\dfrac{c\,\tau_P}{2}$
- Frequency spacing: $\dfrac{1}{\tau_P}$
- Radial-velocity spacing: $\dfrac{\lambda}{2\,\tau_P}$

These ambiguities can produce measurement errors and confusion of multiple targets. However, by choosing τ_P and λ to avoid ambiguities for a target cluster, the cluster can be examined with very high resolution. Pulse-burst waveforms can also be used to reduce the exoatmospheric chaff in the range-radial velocity resolution cell (Sec. 4.7). The ambiguity peaks may be suppressed by using non-uniform pulse spacing, at the expense of the ambiguity regions being wider in both range and radial velocity.

5.2   Moving Target Indication (MTI) and Displaced Phase-Center Array (DPCA)

Clutter from terrain and the sea surface has a velocity spread of a few m/s or less, centered at zero velocity (Sec. 4.5). Moving target indication (MTI), is often used by ground based radars to cancel the clutter without affecting the return from most targets that have non-zero radial velocity. This technique employs coherent processing of two or more pulse returns to create a null region around zero Doppler frequency to reject the clutter spectrum. The displaced phase center array, (DPCA), is an extension of this technique to airborne radars, where the clutter radial velocity and velocity spread is increased by aircraft motion, as discussed later in this section.

Clutter cancellation. A simple two-pulse canceller coherently subtracts successive pulse returns, producing a sharp null at zero velocity. Cascading additional cancellers so that signals from three or more pulses are processed produces a broader null that better cancels clutter having a finite velocity spread, (Sec. 4.5). The MTI improvement factor, defined as the increase in signal-to-clutter ratio (S/C), decreases with increasing clutter-velocity spread, PRF and, wavelength, but is greater for three-pulse cancellers than for two-pulse cancellers, (Fig. 5.2), [36].

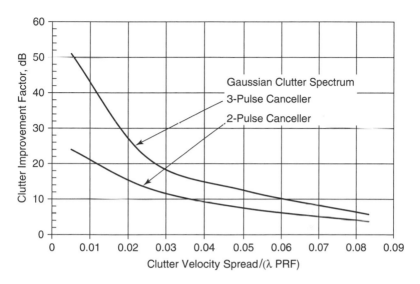

Fig. 5.2 Clutter improvement factor for two- and three-pulse MTI cancellers vs. normalized clutter velocity spread standard deviation (data from [36])

Impact on target signals. Targets having zero or a small radial velocity will be cancelled along with the clutter. Note that this can include high-velocity targets on paths tangential to the radar line-of-sight, (LOS). The smallest radial velocity that will not significantly affect the target signal is called the minimum detectable velocity, MDV. If this is defined as the radial velocity where the target signal power is equal to that of a single pulse without canceling [36]:

$$MDV \approx 0.08\lambda PRF \qquad (5.9)$$

MTI filters also produce nulls at frequencies equal to multiples of the PRF. The corresponding radial velocities are called blind speeds, V_B, and occur at:

$$V_B = \pm \frac{n\,\lambda\,PRF}{2} \qquad (5.10)$$

where n is an integer. The target signal will be below the signal level without MTI for radial velocities ±MDV around each of these blind speeds. These periodic blind speeds may be avoided by using a sufficiently-high PRF, or by varying the PRF during MTI processing.

Displaced phase-center array (DPCA). For airborne and space-based radars having a significant platform velocity, the radial velocity of the clutter at the beam center can be large. Also, the spread of main-beam clutter radial velocities can be significantly larger than the actual clutter velocity spread (Sec. 5.3).

DPCA eliminates this radial-velocity clutter spread by moving the phase center of the radar antenna in the opposite direction of the platform motion between the pulses used for cancellation. This is done by selecting different portions of a longitudinal antenna array on the platform to transmit and receive on successive pulses. The radar PRF is adjusted to the phase-center spacing, d, and the platform velocity, V_P:

$$\text{PRF} = \frac{V_P}{d} \tag{5.11}$$

Since the clutter returns from successive pulses are identical, an MTI processor, can be used to cancel the clutter.

DPCA is a special case of space-time adaptive processing (STAP), discussed in Sec, 5.3. Pulse-Doppler processing (Sec. 5.3) is also used for clutter rejection with airborne and space-based radars.

5.3 Pulse Doppler and Space-Time Adaptive Processing (STAP)

Pulse-Doppler processing is widely used by airborne radars to detect targets in the presence of terrain and sea clutter. A train of pulses is transmitted, and the received signals are processed using a Fourier transform or similar technique, to resolve the signals into a series of narrow spectral bands. These narrow spectral bands, along with good range resolution, can reduce or eliminate the clutter in the target resolution cell, allowing target detection and tracking.

Clutter characteristics. For airborne radars, surface clutter can extend in range to the radar horizon (Sec. 4.4), and in radial velocity from $\pm V_P \cos \phi_D$, where V_P is the platform velocity and ϕ_D is the depression angle (the elevation angle from the platform velocity to the radar LOS).

The radial velocity of the clutter at the beam center, V_C, is (Fig. 5.3):

$$V_C = V_P \cos \phi_A \cos \phi_D \tag{5.12}$$

where ϕ_A is the azimuth angle from the platform velocity to the radar LOS. The clutter velocity spread, V_S, in the radar beam is:

$$V_S = V_P \theta_A \left| \sin \phi_A \right| \cos \phi_D \tag{5.13}$$

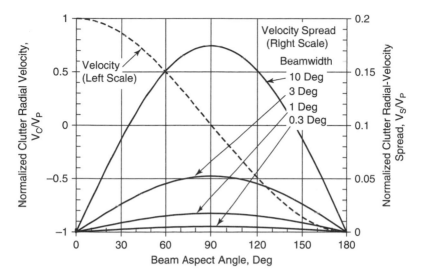

Fig. 5.3 Normalized clutter radial velocity and radial-velocity spread for several azimuth beamwidths (at beam half-power level) vs. aspect angle relative to platform velocity, assuming zero depression angle.

where θ_A is the azimuth beamwidth, and V_S is measured at the half-power level of the azimuth beam; (the null-to-null V_S is approximately twice this value). The minimum detectable velocity, (MDV), for targets relative to main-beam clutter is:

$$\text{MDV} = 0.5\, V_P\, \theta_A\, |\sin\,\phi_A|\,\cos\,\phi_D \qquad (5.14)$$

Clutter returns also enter the radar sidelobes. These can be significant especially with high-PRF waveforms where waveform ambiguities allow clutter returns from short ranges to compete with long-range target signals. The sidelobe clutter radial velocity depends on the azimuth (Eq. 5.13). The sidelobe returns come from 360 degrees and so have a spectral width of $\pm V_P \cos\phi_D$. The range extent of the sidelobe clutter is determined by the radar range resolution (Sec. 4.5).

The clutter return from high depression angles is large, due to both the short range to the terrain and the large values of clutter reflectivity at high grazing angles (Sec. 4.5). This clutter is often called the altitude return, and has a relatively narrow spectral width around zero radial velocity.

Airborne moving target indication (AMTI). Pulse-Doppler waveforms are commonly used by airborne radars for detecting moving aircraft. These waveforms can produce ambiguities in both range and

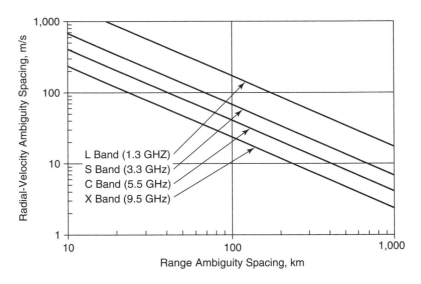

Fig. 5.4 Radial-velocity ambiguity spacing vs. range ambiguity spacing for four radar frequencies

radial velocity. The PRF of the radar pulse train determines the spacing of range ambiguities, R_A, and radial velocity ambiguities, V_A, (Fig. 5.4):

$$R_A = \frac{c}{2 \, \text{PRF}} \tag{5.15}$$

$$V_A = \frac{\lambda \, \text{PRF}}{2} \tag{5.16}$$

Pulse-Doppler radars can employ one of three PRF classes, depending on the target characteristics and the observation geometry (Table 5.2):

- Low PRF is unambiguous in target range, but highly ambiguous in radial velocity. It is used when the target is not in main-beam clutter. This is the case when the target is at high altitude and beyond the radar horizon or at a high enough elevation angle that clutter is not a concern.
- Medium PRF is ambiguous in both range and radial velocity, but the ambiguity spacings usually allow both to be resolved when the target is in track. It is used with targets having ranges and radial velocities that are within the sidelobe clutter region. Sidelobe clutter in the target range-radial velocity resolution cell compete

TABLE 5.2   Pulse Repetition Frequency (PRF) Classes

Low PRF	• Unambiguous in R	$\mathrm{PRF} \leq \dfrac{c}{2\,R_{\mathrm{MAX}}}$
	• Highly ambiguous in V_R	
Medium PRF	• Ambiguous in R and V_R	$\dfrac{c}{2\,R_{\mathrm{MAX}}} < \mathrm{PRF} < \dfrac{2\,V_{R\,\mathrm{MAX}}}{\lambda}$
	• Ambiguities usually can be resolved	
High PRF	• Unambiguous in V_R	$\mathrm{PRF} \geq \dfrac{2\,V_{R\,\mathrm{MAX}}}{\lambda}$
	• Highly ambiguous in R	

with the target, but clutter at other ranges and radial velocities is rejected.

• High PRF is unambiguous in radial velocity, but highly ambiguous in range. It is used for targets having radial velocities that exceed the platform velocity, V_P, as is the case for targets approaching the frontal aspect of the radar platform. These targets are free of clutter in the radial-velocity domain.

Ground moving target indication (GMTI). Surface targets and other slow-moving targets with a clutter background may be observed using medium-PRF pulse-Doppler waveforms. However, the clutter radial-velocity spread, (Eq. 5.13), is usually much greater than the spread of the actual clutter velocity (Sec. 4.5), and the MDV, (Eq. 5.14), often exceeds the velocity of surface targets of interest. Space-time adaptive processing, (STAP), enables observation of targets in main-beam clutter that have radial velocities smaller than pulse-Doppler MDV.

The clutter radial velocity varies with azimuth angle and occupies a diagonal region in the angle-radial velocity cell. Conventional radars separately process target angle (space) using the antenna, and radial velocity (time) using the pulse-Doppler processor. With STAP, the angle and radial velocity are jointly processed, allowing observation of targets that are within the angle-velocity cell but outside the actual clutter velocity spread.

STAP processes each of N pulses in the pulse-Doppler train with each of M horizontal elements in the antenna array. A set of *NM* complex weights is calculated and applied to the signal data for each range-resolution interval. This calculation is computationally intense and involves inversion of matrices with dimensions *NM* × *NM*. Since not all the elements contain useful information, techniques are usually used to reduce the matrix rank and the computational load.

5.4 Synthetic-Aperture Radar (SAR)

Synthetic-aperture radars produce the effect of very-long antenna apertures by using the linear motion of the radar platform, usually an aircraft or satellite. The radar transmits and receives a series of pulses as the radar platform moves along its path. These are coherently processed using a Fourier-transform-like algorithm to produce a very-narrow azimuth beamwidth, usually much smaller than is possible with a real aperture confined to the radar platform (Sec. 2.1). The narrow beamwidth produces small cross-range resolution, often comparable to the radar range resolution (Sec. 5.1), supporting generation of two-dimensional maps of terrain and other surface features.

Synthetic-aperture parameters. The synthetic-aperture length, W_S, is:

$$W_S = V_P t_P \tag{5.17}$$

where V_P is the platform velocity and t_P is the processing time. The beamwidth produced by this aperture is:

$$\theta_S = \frac{\lambda}{2\,W_S\,\cos\varphi} = \frac{\lambda}{2\,V_P\,t_P\,\cos\varphi} \tag{5.18}$$

where φ is the viewing angle off the normal to the platform velocity. (This is half the beamwidth of a real aperture having this dimension because transmission and reception is done at each real antenna location.) The resulting cross-range resolution, ΔD, is:

$$\Delta D = \frac{R\,\lambda}{2\,W_S\,\cos\varphi} = \frac{R\,\lambda}{2\,V_P\,t_P\,\cos\varphi} \tag{5.19}$$

The SAR PRF must be high enough to produce the effect of a filled array to avoid grating lobes:

$$\text{PRF} \geq \frac{4\,V_P}{W} \approx \frac{4\,V_P\,\theta_A}{\lambda} \tag{5.20}$$

where W is the real-aperture length, and θ_A is the real-aperture azimuth beamwidth [37].

The rate of signal phase change during SAR processing determines the cross-range position of targets. Moving targets that have radial velocity different from the terrain will appear offset in angle from their true position. This may be detected and corrected by additional processing when needed.

While conventional SAR generates two-dimensional terrain images, some SARs use two antennas, offset vertically on the radar platform.

Terrain altitude can then be measured using interferometry between the two SAR images.

SAR relies on a moving radar platform. However, SAR techniques can also image targets having rotational motion, either due to actual target rotation or to changes in the observation angle as the target passes the radar or the radar passes the target. This is called inverse synthetic-aperture radar (ISAR), and is further discussed in Sec. 5.5).

The synthetic aperture length, W_S, that can be generated, and hence the cross-range resolution, depends on the complexity of the SAR processing. Three levels of processing are described below in order of increasing complexity.

Doppler beam sharpening (DBS). This technique, also called unfocused Doppler processing, assumes a linear phase progression for successive returns. The synthetic-aperture length must be small enough that the target range changes during processing are small compared with the radar wavelength:

$$W_S \leq \sqrt{R\lambda} \qquad (5.21)$$

The synthetic-aperture beamwidth is limited to:

$$\theta_S \geq \frac{1}{2}\sqrt{\frac{\lambda}{R}} \qquad (5.22)$$

and the cross-range resolution is limited to:

$$\Delta D \geq \frac{\sqrt{R\lambda}}{2} \qquad (5.23)$$

While the cross-range resolution with DBS is usually not useful for imaging, the technique is relatively easy to implement, and can be used to improve the azimuth resolution of closely-spaced targets and improve the azimuth measurement accuracy.

Side-looking SAR. This technique employs signal processing, called focused processing, that corrects for phase changes as the radar passes the target elements. Side-looking SAR generates a continuous image of the terrain as the platform moves along its path, often called a strip map.

The synthetic-aperture length is limited by the azimuth viewing angles for which the target elements remain in the real-aperture beamwidth. For a real-aperture beamwidth of θ_A in the observation direction, φ:

$$W_S \leq \frac{R\,\theta_A}{\cos\varphi} \approx \frac{R\,\lambda}{W\cos\varphi} \qquad (5.24)$$

where W here is the projection in the observation direction of the real-aperture length. The synthetic-aperture beamwidth is limited to:

$$\theta_S \geq \frac{\lambda}{2\,R\,\theta_A} \approx \frac{W}{2\,R} \tag{5.25}$$

and the cross-range resolution is limited to:

$$\Delta D \geq \frac{\lambda}{2\,\theta_A} \approx \frac{W}{2} \tag{5.26}$$

Many dedicated SAR platforms employ antennas directed normal to the flight path, giving values of $\varphi = 0$ and $\cos\varphi = 1$. Multi-function airborne radars often use beams pointed off the normal. In these cases, the synthetic aperture length increases, which would produce narrower beams, but this advantage is lost due to off-broadside beam broadening, which broadens the beam to give the same value as obtained normal to the flight path.

Spotlight SAR. In this mode, the real-aperture beam is steered to remain on a specified target area during the desired SAR processing time, t_P. The signal processing is focused in that it corrects for phase changes of the target elements as the ranges of the elements change. The synthetic aperture length, beamwidth and cross-range resolution are given by Eqs. 5.17, 5.18, and 5.19 respectively. When viewing the target area at an angle off normal, the synthetic-aperture beamwidth and cross-range resolution vary with $\cos\varphi^{-1}$.

For very-long processing time and small range resolution, it may be necessary to correct for motion of target elements between range and cross-range resolution cells. This is called extended coherent processing.

Spotlight SAR is used when better cross-range resolution is required than can be provided by side-looking SAR. This may be the case when a large antenna is used in order to achieve sensitivity. The area imaged is limited by the real-aperture beamwidth. Phased-array radars may generate images of several areas simultaneously by illuminating them in sequence at the required PRF, and even produce the equivalent of strip maps.

5.5   Classification, Discrimination, and Target Identification

Radars can provide information to characterize targets. Such characterization can be considered in three levels (Table 5.3):

- Classification is the process of identifying the target type or class, such as fighter aircraft, helicopter, missile, or truck. This can often

TABLE 5.3 Levels of Target Characterization and Radar Techniques

Characterization Level	**Relevant Radar Techniques**
Classification	Tracking
	RCS measurement
	Length measurement
	Contextual information
Discrimination	RCS and RCS fluctuation measurement
	Range profile
	Spectral signature
	Range-Doppler image
	Precision tracking
Identification	Contextual information
	SSR/IFF transponder responses

be determined from target motion characteristics, and from radar measurements of target size.

- Discrimination is the process of differentiating between targets of interest and other similar objects that may include decoy objects that are made to resemble the targets of interest. This usually involves detailed measurement of radar signatures and motion path parameters.

- Identification is the process of determining the specific identity of a target, such as a specific ship or the flight number of a transport aircraft. Target identification may be accomplished by comparing radar-obtained characteristics with known contextual information, but it often requires target cooperation and the use of transponder responses from the target.

The types of radar data that can support these functions are discussed below.

Radar tracking. The characteristics of a target path determined by radar tracking provide data useful to all these characterization levels:

- Gross measurements of target path can separate targets into major classes such as stationary surface objects, ships, surface vehicles, aircraft, helicopters, and missiles.

- More precise measurements of target altitude, velocity, maneuver, and acceleration can further separate objects within these classes.

For example, fighter aircraft from transport aircraft, slow-moving vehicles like tanks from faster vehicles like trucks and cars.
- Measuring the deceleration of missile targets reentering the atmosphere, or within the atmosphere, allows estimation of their aerodynamic coefficients, which may allow them to be separated from decoys and other objects.

Radar measurements. The measurement of target radar characteristics provides a great deal of information about the configuration of a target:

- Target RCS provides a rough measurement of target size, but must be used with caution since stealth techniques may reduce RCS, and devices like specular reflectors may increase RCS (Sec. 3.1). Radars having dual polarizations provide information on target complexity and the orientation of linear elements, (Sec. 3.1). RCS measurements are subject to bias errors due to atmospheric attenuation and rain (Sec. 4.1 and 4.2).
- Target RCS fluctuations provide information on target size and dynamics. The fluctuation rate at a fixed frequency gives a relationship between size and rotation rate (Eq. 3.3) while measurements of a stable target at different frequencies can infer target size (Eq. 3.4).
- Target radial length and radial range profile can be measured by radars having range resolution better than the target radial length, (Sec. 5.1). This provides direct information on target size and useful information on the target structure. Variation of these measurements with time gives further size information, as well as information on target dynamics.
- Target cross-range dimension and rotational velocity of a target may be found from the spectral spread of the returned signal, Δf:

$$\Delta f = \frac{2\,a\,\omega \sin \gamma}{\lambda} \tag{5.27}$$

where a is the target dimension, ω is the rotational velocity (in radians/second), and γ is the angle between the radar LOS and the target rotational axis. The spectral spread and Doppler-frequency profile within the spread can be measured using a pulse burst waveform having frequency resolution smaller than Δf, and frequency ambiguities separated by more than Δf (Sec. 5.1).

- A range-Doppler image of a rotating target may be generated by producing a Doppler-frequency profile, as described above, for each range resolution interval. Since the Doppler frequency corresponds to the cross-range dimension, this is essentially a two-dimensional target image. The cross-range dimension may be calibrated using Eq. 5.27, if the target rotation rate can be determined, for example by observing target features, and the aspect angle estimated. This inverse synthetic-aperture radar (ISAR) technique uses processing similar to SAR (Sec. 5.4).

Contextual information. The context of radar observations of an object can provide additional information for object characterization:

- The locations of surface vehicles relative to roads and other features can indicate their characteristics. For example, off-road vehicles may be tanks while on-road vehicles are more likely trucks.
- The grouping of surface targets can be reveal their character. Configurations of both moving convoys and fixed installations may indicate their nature and the types of objects employed. The same holds true for shipping convoys, naval formations, and specific ship configurations.
- The point of origin for flight paths and estimates of the destination can reveal the target type and objective. It may be possible to identify a specific object from flight-plan data or information on objects based at the point-of origin. Similar inferences can be made based on ship paths.
- The responses of a target may provide useful information. For example, a target may avoid urban areas or defensive installations. Targets may maneuver to avoid unexpected obstacles such as enemy fighters, or they may respond to external commands. The latter technique has been widely used by air controllers to identify specific transport aircraft.

Secondary surveillance radar (SSR), and identification friend of foe (IFF). In these systems, a ground station transmits interrogation pulses using a rotating antenna, which is usually combined with an air-surveillance radar antenna. Each aircraft carries a transponder that receives and decodes the interrogation, and replies on a different frequency with encoded information, usually including aircraft identification and altitude. The ground station receives and decodes this information, providing positive identification and location of the aircraft (Fig. 5.5).

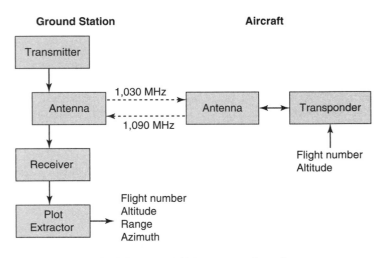

Fig. 5.5   IFF and SSR system configuration

This concept was developed as Identification Friend of Foe (IFF) during World War II to provide secure aircraft identification. The current version is the Mk XII IFF system. It has been adapted for air-traffic control use with the addition of altitude data. It was originally called the Discrete Address Beacon System (DABS). It is now called Secondary Surveillance Radar (SSR) or the Mode S system [38].

The system has the following features:

- Each aircraft must be equipped with a transponder.
- One-way transmissions provide long range with low power and relatively small antennas.
- Separate up-link and down-link frequencies avoid extraneous radar returns including clutter.
- Positive and unambiguous information is provided, including target identification and altitude.
- Target range is provided from the response delay time.
- Target azimuth is provided by monopulse antenna measurement.
- Target altitude is measured by the aircraft, avoiding the need for height measurement by air surveillance radars.
- Target identification code in IFF may be secure to prevent enemy exploitation.

6 | Computation Aids

6.1 Units and Conversion Factors

This book uses the International System of Units (SI), also known as the metric or mks system (for meter, kilogram, second). Factors for converting to and from other common units are given in Table 6.1.

Parameter values are often expressed in units that are multiples of SI units by a factor of ten to some power. These factors, along with the corresponding prefix and its abbreviation, are given in Table 6.2.

TABLE 6.1 Factors for Converting from and to the SI System of Units

Convert This Unit	To This Unit	Multiply By	Inverse Factor
Meters	Yards	1.094	0.9141
Meters	Feet	3.281	0.3048
Meters	Inches	39.37	0.02540
Kilometers	Miles	0.6214	1.609
Kilometers	Nautical miles	0.5400	1.852
Kilometers	Kilo-feet	3.281	0.3048
Meters/second	Kilometers/hour	3.600	0.2778
Meters/second	Miles/hour	2.237	0.4470
Meters/second	Nautical miles/hour (knots)	1.944	0.5145
Kilograms	Pounds	2.205	0.4536
Grams	Ounces	0.03527	28.35
Seconds	Minutes	0.01667	60.00
Seconds	Hours	0.0002778	3600
Radians	Degrees	57.30	0.01745
Joules	Calories	0.2388	4.187
Kilowatts	Horsepower	1.341	0.7457

TABLE 6.2 Measurement Unit Prefixes and Their Corresponding Factors

Prefix	Abbreviation	Factor
Pico	p	10^{-12}
Nano	n	10^{-9}
Micro	μ	10^{-6}
Milli	m	10^{-3}
Centi	c	10^{-2}
Deci	d	10^{-1}
Deka	da	10^{1}
Hecto	h	10^{2}
Kilo	k	10^{3}
Mega	M	10^{6}
Giga	G	10^{9}
Tera	T	10^{12}

Temperatures are specified in SI by Kelvin (K). These can be converted to and from degrees Celsius, (C), and degrees Fahrenheit (F), by:

$$K = C + 273.15 \tag{6.1}$$

$$C = K - 273.15 \tag{6.2}$$

$$K = 273.15 + \frac{5(F - 32)}{9} \tag{6.3}$$

$$F = \frac{9K}{5} - 459.67 \tag{6.4}$$

6.2 Constants

The precise values of constants used in this book are given in Table 6.3.

The velocity of light in the sea-level atmosphere is 2.997×10^8 m/s. A value of 3×10^8 m/s is commonly used both within and outside the atmosphere, as is a value of 1.38×10^{-23} J/K for Boltzmann's constant.

A common room-temperature value often used in radar noise calculations is 290 K. This corresponds to approximately 17 degrees Celsius and 62 degrees Fahrenheit.

TABLE 6.3 Constants for Radar Analysis

Constant	Abbreviation	Value
Velocity of light in a vacuum	c	2.998×10^8 meters/second
Boltzmann's constant	k	1.381×10^{-23} Joules/Kelvin
Mean earth radius		6,371 kilometers
4/3 earth radius	r_E	8,495 kilometers

6.3 Decibels

Power ratios are often expressed in logarithmic form by decibels (dB):

$$dB = 10 \log (\text{power ratio}) \qquad (6.5)$$

$$\text{power ratio} = 10^{\frac{db}{10}} \qquad (6.6)$$

Radar power ratios expressed in decibels include signal-to-noise ratio, antenna gain, and loss factors. Quantities relative to a specific measurement unit are also specified as decibels relative to that unit. Examples are dB relative to a Watt (dBW), dB relative to a square meter, often used for RCS, (dBsm or dBm^2), and dB relative to antenna isotropic gain, (dBI).

Since decibels are logarithms, adding and subtracting them is equivalent to multiplying and dividing the corresponding power ratios, and multiplying decibels by a factor is equivalent to raising the corresponding power ratio to a power equal to the factor (Table 6.4). Thus radar losses

TABLE 6.4 Equivalence of Decibel and Power Ratio Operations

Decibel Operation	Equivalent Power-Ratio Operation
A (in dB) $+$ B (in dB)	A (power ratio) $\times$ B (power ratio)
A (in dB) $-$ B (in dB)	$\dfrac{A \text{ (power ratio)}}{B \text{ (power ratio)}}$
$C \times D$ (in dB)	D (power ratio)^C
$\dfrac{D \text{ (in dB)}}{C}$	D (power ratio)$^{-C}$
E (in dB) $+$ $10n$	E (power ratio) $\times 10^n$
E (in dB) $-$ $10n$	$\dfrac{E \text{ (power ratio)}}{10^n}$

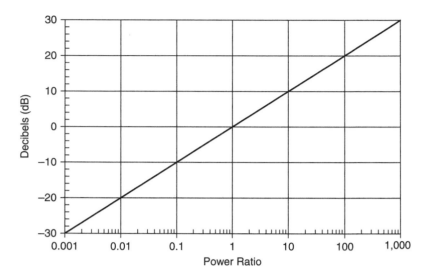

Fig. 6.1   Decibel values corresponding to power ratio values

that are multiplicative can be calculated by summing their dB values.

Fig. 6.1 shows the relationship between power ratios and decibels. Tables 6.5 and 6.6 give decibel values for common power ratios and power ratios for common decibel values respectively. Note that multiplying a

TABLE 6.5   Power Ratios and Their Decibel Values

Power Ratio	Decibels	Power Ratio	Decibels
1	0	0.1	−10
2	3.010	0.2	−6.989
3	4.771	0.3	−5.229
4	6.021	0.4	−3.970
5	6.990	0.5	−3.010
6	7.782	0.6	−2.218
7	8.451	0.7	−1.549
8	9.031	0.8	−0.9691
9	9.542	0.9	−0.4576
10	10	1	0

TABLE 6.6 Decibels and Their Power-Ratios

Decibels	Power Ratio	Decibels	Power Ratio
0	1	0	1
1	1.259	-1	0.7943
2	1.585	-2	0.6310
3	1.995	-3	0.5012
4	2.512	-4	0.3981
5	3.162	-5	0.3162
6	3.981	-6	0.2512
7	5.012	-7	0.1995
8	6.310	-8	0.1585
9	7.943	-9	0.1259
10	10	-10	0.1

power ratio by 10^n adds $10n$ to the decibel value, and dividing a power ratio by 10^n subtracts $10n$ from the decibel value, and conversely (Table 6.4).

Symbols

A	Effective antenna aperture area, m^2
a	Target dimension, m
A_A	Physical antenna aperture area, m^2
a_A	Two-way atmospheric loss, dB/km
A_C	Surface clutter area, m^2
A_{CR}	Projected area of corner reflector, m^2
A_E	Array element effective aperture area, m^2
A_P	Flat plate area, m^2
A_R	Receive antenna effective aperture area, m^2
a_R	Two-way rain loss, dB/km
A_φ	Array effective aperture area at scan angle φ, m^2
B	Signal bandwidth, Hz
B_J	Jammer bandwidth, Hz
B_R	Receiver bandwidth, Hz
B_S	Subpulse bandwidth, Hz
C	Clutter or chaff return power, W
c	Velocity of light, 3×10^8 m/s
CR	Cancellation ratio
D	Antenna directivity; cross-range dimension, m
d	Array element spacing, m; DPCA phase center spacing, m
DC	Transmitter duty cycle
E	Pulse or waveform energy, J
E_{MAX}	Maximum pulse energy, J
ERP	Jammer effective radiated power, W
f	Frequency, Hz
f_D	Doppler-frequency shift, Hz
f_R	Doppler-frequency resolution, Hz
F_R	Receiver noise figure or noise factor
G	Antenna gain
G_E	Array element gain
G_J	Jammer antenna gain
G_R	Receive antenna gain
G_T	Transmit antenna gain
G_φ	Array gain at scan angle φ
h_R	Radar altitude, m
h_T	Target altitude, m

J	Jammer signal power in radar, W
k	Boltzmann's constant, 1.38×10^{-23} J/K
k_A	Antenna beamwidth coefficient
L	Radar system losses
l	Rain or atmosphere attenuation path length, m
L_A	Total two-way atmospheric loss
L_{BS}	Beam-shape loss
L_D	Detection loss
L_E	Antenna aperture-efficiency loss
L_J	Jammer loss
L_M	Receive microwave loss
L_O	Antenna ohmic loss
L_R	Total two-way rain loss
L_S	Search losses
L_T	Transmit losses
N	Noise power, W; atmospheric refractivity
n	Number of pulses integrated; number of pulses that observe target; number of pulses used in measurement; refractive index
n_B	Number of beams in search pattern
n_C	Number of chaff dipoles
n_E	Number of array elements
n_M	Number of phased-array modules
n_S	Number of subpulses
P_A	Average transmitter power, W
P_{AM}	Module average power, W
PC	Signal gain from pulse compression
P_D	Probability of detection
P_{DO}	Detection probability for single observation in cumulative detection
P_{FA}	Probability of false alarm
P_{FAO}	False-alarm probability for single observation in cumulative detection
P_J	Jammer power, W
P_P	Peak transmitter power, W
P_{PM}	Module peak power, W
P_R	Receiver noise power, W
PRF	Pulse repetition frequency Hz
PRI	Pulse repetition interval, s

P_S	Prime power supplied to transmitter, W
r	Rainfall rate, mm/hr
R	Range from radar to target, m
R_A	Assured acquisition range, m; range ambiguity spacing, m
R_D	Detection range, m
r_E	4/3 earth radius, 8,495 km
R_F	Far-field range, m
r_{FA}	False-alarm rate, Hz
R_H	Horizon range, m
R_M	Minimum range constraint, m
R_{MAX}	Maximum target range, m
R_R	Range from target to receive antenna, m
R_S	Cylinder radius, m
R_T	Range from transmit antenna to target, m; shortest range to target in barrier search, m
S	Signal power, W
S/C	Signal-to-clutter ratio; signal-to-chaff ratio
S/J	Signal-to-jammer ratio
S/N	Signal-to-noise ratio
SL	Sidelobe level relative to antenna gain
SLI	Sidelobe level relative to isotropic gain
t	Round-trip propagation time, s
T_A	Sky temperature, K
t_{FA}	Time between false alarms, s
t_M	Measurement time duration, s
t_R	Antenna rotation period, s
T_R	Receiver noise temperature, K
t_R	Time resolution or compressed pulse duration, s
T_{RM}	Module receiver noise temperature, K
t_S	Search time, s
T_S	System noise temperature, K
V	Target velocity, m/s
V_A	Radial-velocity ambiguity spacing, m/s
V_C	Clutter or chaff volume in resolution cell, m^3
$V_{R\,\mathrm{MAX}}$	Maximum target radial velocity, m/s
V_R	Target radial velocity, m/s
V_T	Target tangential velocity in barrier search, m/s; total chaff volume, m^3
W	Antenna dimension, m; real-aperture length, m

W_C	Chaff weight, kg
W_S	Flat plate dimension, m; antenna subarray dimension, m; synthetic-aperture length, m
Y_T	Threshold level normalized to mean noise power
α	Angle between target velocity vector and radar line-of-sight, radians; tracking filter position parameter
α_R	Angle between target velocity vector and receive line-of-sight, radians
α_T	Angle between target velocity vector and transmit line-of-sight, radians
β	Tracking filter velocity parameter; bistatic angle, radians
ΔD	Cross-range resolution, m
Δf	Change in frequency for independent observation, Hz; signal spectral spread, Hz
ΔR	Range resolution, m
δR	Direct and multipath reflected path-length difference, m
ΔV	Radial-velocity resolution, m/s
$\Delta \alpha$	Change in aspect angle for independent observation, radians
$\Delta \phi$	Elevation-angle separation of lobes from multipath, radians
ϕ_A	Azimuth angle between platform velocity and radar line-of-sight, radians
ϕ_D	Depression angle, radians
ϕ_E	Elevation angle dimension of barrier search, radians
γ	Grazing angle, radians; angle between target rotational axis and radar line-of sight, radians; tracking filter acceleration parameter
η	Volume clutter reflectivity, m^{-1}
η_T	Transmitter efficiency
φ	Scan angle off array broadside, radians; viewing angle off the normal to platform velocity vector, radians
φ_M	Maximum array scan angle, radians
λ	Wavelength, m
θ	Antenna beamwidth, radians
θ_A	Azimuth beamwidth, radians
θ_E	Elevation beamwidth, radians
θ_S	Synthetic-aperture beamwidth, radians
θ_X	Antenna beamwidth in X plane (normal to the Y plane), radians

θ_Y	Antenna beamwidth in Y plane (vertical), radians
θ_φ	Array beamwidth at scan angle ϕ, radians
σ	Radar cross section, m^2; standard deviation of measurement error
σ^0	Surface clutter reflectivity
σ_A	Standard deviation of angle measurement error, radians
σ_{AV}	Average radar cross section value, m^2
σ_C	Clutter radar cross section, m^2; total chaff radar cross section, m^2
σ_D	Standard deviation of cross-range measurement error, m
σ_R	Standard deviation of range measurement error, m
σ_V	Standard deviation of radial-velocity measurement error, m/s
τ	Waveform duration, s
τ_{MAX}	Maximum transmitter pulse duration, s
τ_P	Subpulse time separation, s
τ_S	Subpulse duration, s
ω	Target rotational velocity, radians/s
ψ	Angle from antenna main beam, radians
ψ_S	Search solid angle, radians^2

Glossary

AGC	Automatic gain control
AMTI	Airborne moving target indication
CFAR	Constant false-alarm rate
CM	Countermeasure
COHO	Coherent local oscillator
CW	Continuous wave
DABS	Discrete Address Beacon System
DBS	Doppler beam sharpening
DPCA	Displaced phase-center array
ECCM	Electronic counter-countermeasure
ECM	Electronic countermeasure
EHF	Extremely-high frequency
ERP	Effective radiated power
ESJ	Escort jammer
EW	Electronic warfare
FFOV	Full field-of-view
FFT	Fast Fourier transform
FIR	Finite impulse response
FMCW	Frequency-modulated continuous wave
GaAsFET	Gallium arsenide field-effect transistor
GMTI	Ground moving target indication
GaN	Gallium nitride (transistor)
HEMT	High electron mobility transistor
HF	High frequency
IF	Intermediate frequency
IFF	Identification friend or foe
ISAR	Inverse synthetic aperture radar
ITU	International Telecommunications Union
LC	Left circular polarization
LFOV	Limited field-of-view
LNA	Low noise amplifier
LOS	Line of sight
MDV	Minimum detectible velocity
Mks	Meter-kilogram-second system of units
MLJ	Mainlobe jammer

MTI	Moving target indication
OPS	Operations per second
OTH	Over the horizon
PDF	Probability density function
PRF	Pulse repetition frequency
PRI	Pulse repetition interval
RAM	Radar absorbing material
RC	Right circular polarization
RCS	Radar cross section
RF	Radio frequency
Rss	Root sum square
SAR	Synthetic aperture radar
SAW	Surface-acoustic wave
SHF	Super-high frequency
SI	International System of Units
SLB	Sidelobe blanker
SLC	Sidelobe canceller
SLJ	Sidelobe jammer
SNR	Signal-to-noise ratio
SOJ	Stand-off jammer
SSR	Secondary surveillance radar
STALO	Stable local oscillator
STAP	Space-time adaptive processing
STC	Sensitivity time control
T/R	Transmit-receive
TWS	Track while scan
TWT	Traveling-wave tube
UHF	Ultra-high frequency
VHF	Very-high frequency

References

[1] Curry, G. R., *Pocket Radar Guide,* Scitech Publishing, Raleigh, NC, 2010.

[2] Richards, M. A., Scheer, J. A., and Holm, W. A., *Principles of Modern Radar,* Scitech Publishing, Raleigh, NC, 2010.

[3] Skonlik, M. I., ed., *Radar Handbook,* McGraw Hill, New York, 2008.

[4] Stimson, G. W., *Introduction to Airborne Radar,* Scitech Publishing, Raleigh, NC, 1998.

[5] Curry, G. R., *Radar System Performance Modeling, 2nd ed.,* Artech House, Norwood, MA, 2005.

[6] Barton, D. K., *Radar System Analysis and Modeling,* Artech House, Norwood, MA, 2004.

[7] *IEEE Standard Radar Definitions,* IEEE Std 686-2008, The Institute of Electrical and Electronic Engineers, New York, 2008.

[8] *IEEE Standard Letter Designations for Radar Frequency Bands,* IEEE Std 521-2002, The Institute of Electrical and Electronic Engineers, New York, 2003.

[9] Scheer, J. A. and Holm, W. A., "Introduction and Radar Overview", Chapter 1 in *Principles of Modern Radar,* Ed. Richards, M. A., Scheer, J. A., and Holm, W. A., Scitech Publishing, Raleigh, NC, 2010.

[10] Barton, D. K. and Ward, H. R., *Handbook of Radar Measurement,* Artech House, Norwood, MA, 1984.

[11] Wallace, T. V., Jost, R. J., and Schmid, P. E., "Radar Transmitters", Chapter 10 in *Principles of Modern Radar,* Ed. Richards, M. A., Scheer, J. A., and Holm, W. A., Scitech Publishing, Raleigh, NC, 2010.

[12] Weil, T. A. and Skolnik, M., "The Radar Transmitter", Chapter 10 in *Radar Handbook, 3rd ed.,* Ed. Skolnik, M. I., McGraw Hill, New York, 2008.

[13] Bruder, J. A., "Radar Receivers", Chapter 11 in *Principles of Modern Radar,* Ed. Richards, M. A., Scheer, J. A., and Holm, W. A., Scitech Publishing, Raleigh, NC, 2010.

[14] Taylor, J. W., "Receivers", Ch. 3 in *Radar Handbook 2nd ed.,* Ed. Skolnik, M. I., McGraw Hill, New York, 1990.

[15] Borkowski, M. T., "Solid-State Transmitters", Chapter 11 in *Radar Handbook, 3rd ed.,* Ed. Skolnik, M. I., McGraw Hill, New York, 2008.

[16] Richards, M. A., "The Radar Signal Processor", Chapter 13 in *Principles of Modern Radar,* Ed. Richards, M. A., Scheer, J. A., and Holm, W. A., Scitech Publishing, Raleigh, NC, 2010.

[17] Shaeffer, J. F., "Target Reflectivity", Chapter 6 in *Principles of Modern Radar,* Ed. Richards, M. A., Scheer, J. A., and Holm, W. A., Scitech Publishing, Raleigh, NC, 2010.

[18] Richards, M. A., "Target Fluctuation Models", Chapter 7 in *Principles of Modern Radar,* Ed. Richards, M. A., Scheer, J. A., and Holm, W. A., Scitech Publishing, Raleigh, NC, 2010.

[19] Ruck, G. T., "Planar Surfaces", Chapter 7 in *Radar Cross Section Handbook,* Ed. Ruck, G. T., Plenum Press, New York, 1970.

[20] Barrick, D. E., "Cylinders", Chapter 4 in *Radar Cross Section Handbook,* Ed. Ruck, G. T., Plenum Press, New York, 1970.

[21] Ruck, G. T., "Complex Bodies", Chapter 8 in *Radar Cross Section Handbook,* Ed. Ruck, G. T., Plenum Press, New York, 1970.

[22] Skolnik, M. I., *Introduction to Radar Systems,* McGraw Hill, New York, 2002.

[23] Richards, M. A., "Threshold Detection of Radar Targets", Chapter 15 in *Principles of Modern Radar,* Ed. Richards, M. A., Scheer, J. A., and Holm, W. A., Scitech Publishing, Raleigh, NC, 2010.

[24] Frank, J. and Richards, J. D., "Phased Array Radar Antennas", Chapter 13 in *Radar Handbook, 3rd ed.,* Ed. Skolnik, M. I., McGraw Hill, New York, 2008.

[25] Blair, W. D., "Radar Tracking Algorithms", Chapter 19 in *Principles of Modern Radar,* Ed. Richards, M. A., Scheer, J. A., and Holm, W. A., Scitech Publishing, Raleigh, NC, 2010.

[26] Bath, W. G. and Trunk, G. V., "Automatic Detection, Tracking, and Sensor Integration", Chapter 7 in *Radar Handbook, 3rd ed.,* Ed. Skolnik, M. I., McGraw Hill, New York, 2008.

[27] Blake, L. V., *Radar Range-Performance Analysis,* Artech House, Norwood, MA, 1986.

[28] Morchin, W., *Radar Engineer's Sourcebook,* Artech House, Norwood, MA, 1993.

[29] Crane, R. K., *Electromagnetic Wave Propagation Through Rain,* Wiley & Sons, New York, 1996.

[30] Currie, N. C., "Characteristics of Clutter", Chapter 5 in *Principles of Modern Radar,* Ed. Richards, M. A., Scheer, J. A., and Holm, W. A., Scitech Publishing, Raleigh, NC, 2010.

[31] Wetzel, L. B., "Sea Clutter", Chapter 15 in *Radar Handbook, 3rd ed.,* Ed. Skolnik, M. I., McGraw Hill, New York, 2008.

[32] Moore, R. K., "Ground Echo", Chapter 16 in *Radar Handbook, 3rd ed.,* Ed. Skolnik, M. I., McGraw Hill, New York, 2008.

[33] Millman, G. H., "Atmospheric Effects on Radio Wave Propagation", Ch. 1 in *Modern Radar,* Ed. Berkowitz, R. S., Wiley, New York, 1965.

[34] Keel, B. M., "Fundamentals of Pulse Compression Waveforms", Chapter 20 in *Principles of Modern Radar,* Ed. Richards, M. A., Scheer, J. A., and Holm, W. A., Scitech Publishing, Raleigh, NC, 2010.

[35] Deley, G. W, "Waveform Design", Chapter 3 in *Radar Handbook,* Ed. Skolnik, M. I., McGraw Hill, New York, 1970.

[36] Richards, M. A., "Doppler Processing", Chapter 17 in *Principles of Modern Radar,* Ed. Richards, M. A., Scheer, J. A., and Holm, W. A., Scitech Publishing, Raleigh, NC, 2010.

[37] Showman, G. A., "An Overview of Radar Imaging", Chapter 21 in *Principles of Modern Radar,* Ed. Richards, M. A., Scheer, J. A., and Holm, W. A., Scitech Publishing, Raleigh, NC, 2010.

[38] Stevens, M. C., *Secondary Surveillance Radar,* Artech House, Norwood, MA, 1988.

Index

Notes

Notes